edited by
**ELIZABETH ELLSWORTH**
**+ JAMIE KRUSE**

# MAKING THE
# GEOLOGIC
# NOW

## RESPONSES TO MATERIAL CONDITIONS OF CONTEMPORARY LIFE

punctum books ✳ brooklyn, ny

First published in 2013 by
punctum books
Brooklyn, New York

ISBN-13: 978-0-6157663-6-2

Punctum Books is an open-access and print-on-demand independent publisher dedicated to radically creative modes of intellectual inquiry and writing across a whimsical para-humanities assemblage. We seek to pierce and disturb the wednesdayish, business-as-usual protocols of both the generic university studium and its individual cells or holding tanks. We solicit and pimp quixotic, sagely mad engagements with textual and visual thought-bodies. This is a space for the imp-orphans of your thought and pen, an ale-serving church for little vagabonds.

VISIT GEOLOGICNOW.COM TO JOIN DISCUSSIONS AND POST SIGHTINGS OF THE *GEOLOGIC NOW*.

Image on Front Cover: from the Feasibility Project (Rachel, NV), smudge studio 2009
Image on Back Cover: *An unclassified metallic spherule (possible micrometeorite) on the tip of my index finger.* Image Courtesy: Ryan Thompson, 2012
Image on Frontispiece: Shiprock Uranium Disposal Cell, Shiprock, NM, smudge studio 2012

# CONTENTS

# SIGNALS

SCALES OF FORCE AND CHANGE_ CONFIGURING RAPID ADAPTATION_ GRAPPLING WITH THE POTENCY_ EVENTSCAPES_ STREAMING LANDSCAPES _ MEETING FORCES LARGER THAN OURSELVES_ LIVING THE GEOLOGIC _ IN THE INTEREST OF TIME _ FLARING UP: SPACE WEATHER AND THE BULK POWER SUPPLY_ BURNING THROUGH THE EONS_ GEO-COSMOLOGICAL CONVERGENCE_ THERE IS NO ZERO: CONTINUOUS REMIX IN THE GEOLOGIC CITY_ ANTHROPOCENE GOES MAINSTREAM_ URBAN INFRASTRUCTURE AS GEOLOGIC MATERIALITY IN MOTION_ DESIGN AND EXISTENTIAL RISK_ WE ARE INSTANTANEOUS WITH THE PLEISTOCENE_ WHEN GEOLOGY GETS PERSONAL_ DESIGN FOR INFINITE QUARANTINE_.

# FROM THE EDGE

blog post titles excerpted from Friends of the Pleistocene (fopnews.wordpress.com) 2010-2012.

# INTRODUCTION

Elizabeth Ellsworth and Jamie Kruse

There was never a time when human agency was anything other than an interfolding network of humanity and nonhumanity; today this mingling has become harder to ignore.

—Jane Bennett, *Vibrant Matter: A Political Ecology of Things*

## EVIDENCE: MAKING A GEOLOGIC TURN IN CULTURAL AWARENESS

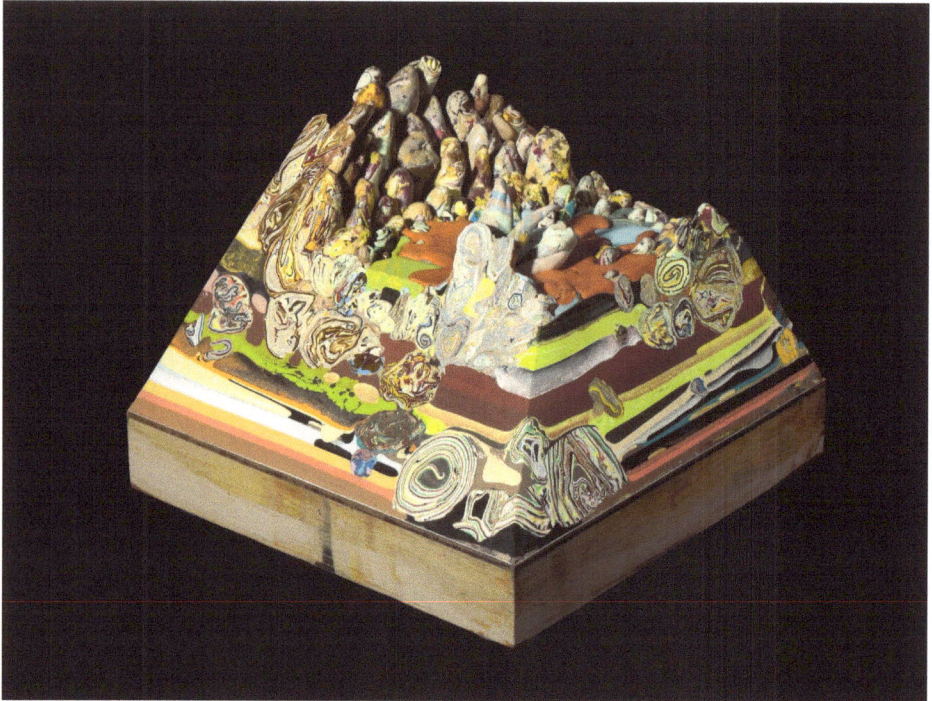

Laura Moriarty, *Subduction into Trench*, 2010, Encaustic on panel

Until recently, the word "geologic" conjured meanings and associations that referred simply and directly to the science of geology—the study of the origin, history, and structures of the earth. But that seems to be changing. Something is happening to the ways that people are now taking up "the geologic."

Contemporary artists, popular culture producers, speculative architects, scientists and philosophers are adding new layers of cultural meaning and aesthetic sensation to the geologic. It is as if recent events and developments are making geologic realities sense-able with new physical intensity and from new angles of thought as a situation that we live within, not simply as something "out there" that we study.

Humans seem to be sensing, in new ways, that the forces and materials of the earth are not only subjects of scientific inquiry—they have also become conditions of daily life.

*The idea for this book came from our sense that there is an increasingly widespread turn toward the geologic as source of explanation, motivation, and inspiration for cultural and aesthetic responses to conditions of the present moment.* Geologic topics and themes are underscoring daily experience in ways that are stark and arresting. Deep time is beginning to have applied, material meaning for non-specialists. Not that long ago, mineral and fossil resources available for human exploitation seemed to be infinite. No nuclear waste needed to be stored. Carbon emissions didn't exist. And the oceans contained zero tons of plastic. Two hundred years ago, fears about what the planet's material state might be like in 1000 years would have seemed misplaced. No longer. In the face of accelerating planetary-scale change, much of it traceable to human activity, today's humans are confronted by realizations that life on earth hasn't been like "this" nor has it looked like "us" for very long at all. In the short 18 months that we've been working on this book, the word "Anthropocene" has gone from an obscure, scientific term that an NPR interviewer had to ask a geologist how to pronounce, to a word that now appears in headlines and returns over half a million Google citations.

Daybreak Utah, smudge studio 2010, from *Below the Line*

We are intrigued by what is arguably a growing recognition that the geologic, both as a material dynamic and as a cultural preoccupation, shapes the "now" in ever more direct and urgent ways. The existence, effects, and nature of earth dynamics were once the specialized interest of scientists and infrastructure designers. Today, they are topics of breaking news about *tectonic plate movements, travel-disrupting volcanic eruptions, deep time, slow accumulations and*

courtesy of William Lamson, *A Line Describing the Sun*, Lake Harper, CA, 2010

*metamorphoses of the world's materiality, erosion and displacement of landforms, dramatic earth reshaping events, and geo-bio interactions.* These are forces to be reckoned with existentially, creatively, conceptually, and pragmatically as humans work to meet the fact that not only is our species increasingly vulnerable to the geologic, we also have become agents of planetary geologic change.

At a conference inspired by the premise of *Making the Geologic Now*, Seth Denizen, one of the contributors to this book, offered stark graphic representations of the accelerating speed of material change on planet earth. On graph after graph (human population growth, numbers of cell phones, ocean acidification), the recent accelerations of change have turned what once appeared as lines with slight slopes upward into near vertical spikes. As Denizen put it, at the points in time and speed that we now are approaching:

> The world becomes defined not by a time, but by a speed. ***This is the point at which the world can no longer be merely an extension of our own, a difference in degree, but rather something which takes on a difference in kind: another sea, another wind, another world at right angles to our own.***

In the wake of Denizen's presentation, it was hard not to suspect that speeds of change in material realities of life on the planet are outpacing our ways of knowing. Most everything we (Western-encultured) humans think we know about living on Earth, and most everything we thought was useful for life here, was invented by ancestors who lived during times when the "hockey stick"-shaped graphs depicting accelerating change were relatively flat. But, as Denizen points out, we're now living at right angles with that (former) world. We're now living on a qualitatively different planet. While that doesn't necessarily mean that human knowledge must start over from here, it does seem to suggest that we need to rethink, reconfigure, and reinvent much of "what we know" from an entirely different angle (the vertical, accelerating rise). And quickly.

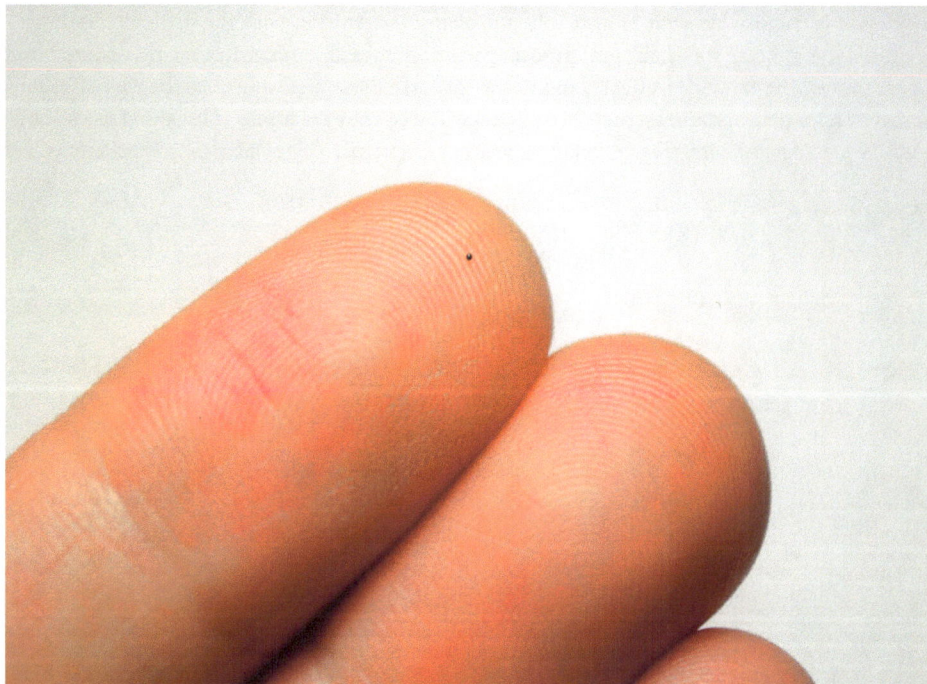

*An unclassified metallic spherule (possible micrometeorite) on the tip of my index finger, Ryan Thompson, 2012*

However, the contributors to this collection show that the urgency of our response can be coupled with complex, multi-leveled ways of addressing change. Some of them are engaged in practices that are designed to take time, pay close attention, take note. They take the rates of changes occurring around them as reason to pause and turn practices of experimental thinking, design, and art towards these changes, in order to respond. Some have traced flows of human-altered geologic materials around the globe (Takeda). Others have made dynamic visual materializations of rates of change (Kavanaugh and Nguyen); noted how life here on earth is always subject to larger, unknowable forces from "outside" our own atmosphere (Thompson); and documented, indirectly, evidence that we humans are "recent" arrivals on this planet who are surrounded by species that have adapted to incredibly varied eco-systems for almost unimaginable spans of time (Sussman). Creators of the works featured here are attending to rates of change with a calm perseverance. *Making the Geologic Now* showcases contributors' abilities to site themselves at emergent material phenomena, invent new practices there, and attempt to imagine, gain, and share new skills capable of meeting urgencies of the moment.

*Shimpei Takeda at Shioyasaki Lighthouse in Iwaki-city, Fukushima, photo by Keisuke Hiei, from Trace — cameraless records of radioactive contamination, Shimpei Takeda, 2012*

New sorts of inventive thinking and making are now possible, and called for, in response to new material situations of daily life. A number of artists, designers, and philosophers of change and emergence are deliberately situating their aesthetic work and experimental thinking *within the geologic as a condition of our present time*. Some contemporary artists locate their bodies and imaginations within jostling and unstable physical, social, political, and economic situations that arise from and act back upon the earth's materialities, forces and events. The result is an increase in aesthetic works that explore and creatively respond to the geologic depth of "now." In the past few years, for example, galleries, museums, and public art projects (including documentary films and blogs) have presented works that do not simply take up the geologic as a theme—they also activate formats, methods, models, ideas, and aesthetic experience in ways that seek to recalibrate "the human" in relation to "the geologic." In 2010, *Rising Currents*, a major exhibition by New York's Museum of Modern Art, traced the co-shaping of geoforms and cultural forms. The exhibited works employed art, architecture, design, science, and city planning to address how the geologic force and event of a rise in global sea levels will affect life in New York City. The same year at the Cooper Union's 41 Cooper Gallery, a group exhibition entitled *The Crude and the Rare* commissioned artists to creatively respond to rare earth materials and offer aesthetic and speculative activations of geologic materiality. In 2011, the Whitney's exhibition of Karthik Pandian's *Unearth* examined relationships between ancient peoples and geomorphology. With *Into Eternity* (2010), film director and conceptual artist Michael Madsen contemplates social, economic, biological, technological, and even spiritual significances of Onkalo, the world's first deep earth repository for high level nuclear waste. Harvard professors Peter Galison and Robb Moss are currently co-producing a documentary, *Nuclear Underground*, that explores the complex set of environmental, political, and moral debates entangled within the problem of nuclear waste disposal in the United States. It will focus on the Waste Isolation Pilot Plant (WIPP) in Carlsbad, New Mexico. In a 2010 show at the Pace Gallery entitled *The Day After*, photographer Hiroshi Sugimoto displayed photograms that resulted when he used electrical sparks and film to create a situation that he compared to "the first meteorite hitting the Earth." The exhibition also displayed fossils and meteorites from his collections. Brian Eno's 2010 album, *Small Craft on a Milk Sea*, includes tracks titled "*Late Anthropocene*" and "*Paleosonic*," and employs music to imagine and articulate meanings and forces of deep time. A number of emergent, hybrid practices, such as the exhibitions and "tours" of the Center for Land Use Interpretation and Lucy Raven's photographic animation *China Town*, are now combining landscape photography, land use interpretation, and public field expeditions into new forms of public art.

courtesy of William Lamson, *A Line Describing the Sun*, 2010

In a related gesture, some contemporary philosophers are experimenting with concepts that take up the geologic as metaphor and model. It seems that the understanding of earth processes can offer inspiration for how we might think about the qualitatively different ways that humans are now living on planet earth. Daily experiences of what has preoccupied philosophers for generations: space, time, matter, and change, are being dramatically altered by new predicaments of speed, scale, flow, and density. Some philosophers are reconsidering and revising assumptions they inherited from Western thought as they bring to their projects new appreciation for the depth and consequences and how humans comingle with the nonhuman.

For example, in *Vibrant Matter: A Political Ecology of Things,*[1] Jane Bennett offers compelling arguments for why and how we might (and even should) approach and understand things geologic as vital forces and active agents in daily life. Drawing on Latour, she points out that "modern selves are feeling increasingly entangled, cosmically, biotechnically, medially, virally, pharmacologically, with nonhuman nature. Nature has always mixed it up with self and society, but lately this comingling has intensified and become harder to ignore" (115). Bennett describes the contemporary moment as "a time when interactions between human, viral, animal, and technological bodies are becoming more and more intense" (108). We have arrived at a moment, she asserts, in which human culture is "inextricably enmeshed with vibrant, nonhuman agencies" (108). Among such nonhuman agents that directly shape daily life, she counts food: "eating constitutes a series of mutual transformations between human and nonhuman materials" (40). She also counts the geologic, and identifies it as a key participant in the effectivity of the electric grid, which she depicts as a "volatile mix of coal, sweat, electromagnetic fields, computer programs, electron streams, profit motives, heat, lifestyles, nuclear fuel" (25). When the grid's complex assemblage of human and nonhuman agents "goes down" or blacks out, it does so with an agency that is "distributed along a continuum, extrudes from multiple sites or many loci—from a quirky electron flow and a spontaneous fire to members of Congress who have a neoliberal faith in market self regulation" (28).

from *Trinity Pilgrimage,* Bryan M. Wilson, 2009

By including geologic material in the realm of the world's "actants," Bennett urges us to break with modern habits of assuming and behaving as if the matter that composes the earth is passive stuff, raw, brute, or inert (vi). Many of us have been encultured to view humans as the active agents on the planet, while things, technologies, and "stuff" are relegated to the status of passive recipients of our designs and desires. But, as Bennett makes clear through examples ranging from stem cells to trash to fish oils, humans are able to act intentionally and

---

1  Jane Bennett, *Vibrant Matter: A Political Ecology of Things* (Durham: Duke University Press, 2010), hereafter referred to parenthetically by page number.

effectively in the world only "if accompanied by a vast entourage of nonhumans" (108).We do not act or move along clearly delineated pathways of human and nonhuman, biologic and geologic, inside and outside, self and other. We operate, instead, within complex "interstitial field[s] of nonpersonal, ahuman forces, flows, tendencies, and trajectories." Bennett suggests that it would be advisable, productive, and even ethical for contemporary philosophers to "theorize a kind of geological affect or material vitality" in recognition of the "shimmering, potentially violent vitality intrinsic to matter" (61).

Such ideas give what used to be defined as the study of the origin, history, and structure of Earth new meanings, questions, and approaches. Of course, "geology" continues to reference rocks, tectonics, and bare forces of our planet, including deep time. But it is taking up new associations as people struggle to understand and meet new and unprecedented material realities of Earth and life on Earth.

To begin to image an expanded field of geology beyond the realm of rocks, we experimented with a passage from Bruno Latour's *Reassembling the Social: An Introduction to Actor-Network Theory*. We took the liberty of appropriating Latour's effort to redirect "sociology" into the messy realm of "the social" (which is always entangled in associations). In the following passage by Latour, we substituted "geology" for "sociology" and "the geologic" for "the social," in order to explore what becomes thinkable when we redirect "geology" into the realm of "the geologic":

> Even though most geologists would prefer to call "the geologic" a homogeneous thing, it's perfectly acceptable to designate by the same word a trail of *associations* between heterogeneous elements. Since in both cases the word retains the same origin—from Latin root geo—it is possible to remain faithful to the original intuitions of the geologic sciences by redefining geology not as the "science of the geologic," but as the *tracing of associations*. In this meaning of the adjective, geologic does not designate a thing among other things, like a black sheep among white sheep, but a *type of connection* between things that are not themselves geologic.[2]

Sudbury, Ontario, smudge studio 2011

2 Experimental reading of Bruno Latour, adapted from *Reassembling the Social: An Introduction to Actor-Network Theory* (Oxford: Oxford University Press, 2005), 5.

# AT SOME POINT DURING THE MEAL WE REALIZED CESIUM LEVELS HAD BEEN POSTED ON THE LUNCH BOARD.

image smudge studio

We think this speculative substitution of terms produces a lively outcome—one that springs geology into the realms of everyday actions, movements, and associations among humans and nonhumans. Thinking this way invites us to "devise new procedures, technologies, and regimes of perception that enable us to consult nonhumans more closely, or to listen and respond more carefully to their outbreaks, objections, testimonies and propositions" (Bennett, 108).

It's becoming more and more difficult to deny that nonhuman aspects of "the geologic" have been objecting, testifying, proposing, and breaking out. Or that they have been doing so with growing intensity and human consequence. The densities and locations of human habitation on Earth, especially in highly technologized urban areas, make humans vulnerable to and aware of geologic forces and change in new ways. When looking for clues about what might be propelling contemporary cultural producers and thinkers to make a geologic turn, recent "natural" and human-made events, some unprecedented in scale and consequence, are obvious suspects. In our call for contributions to *Making the Geologic Now*, we chronicled some of the recent incursions that geologic forces have made into human activities and affairs. Of course, these now include the March 11, 2011, catastrophic, ad hoc assemblage of a 9.0 earthquake, tsunamis of over 70 feet, and three meltdowns in the Fukushima Daiichi nuclear power plants that occurred in Japan. Relatively inchoate senses of "the geologic as a condition of contemporary life" circulating on March 11, 2011 became overwhelmingly palpable and sense-able. In the wake of that day's events, it seems impossible to downplay any longer the reality that assemblages of human technologies and the earth's dynamic geo-materialities can, and do, act back upon us in wildly unpredictable ways. What continues to unfold in Fukushima makes it soberingly clear that the effects of our tools, designs and desires ramify into deep geologic futures. And that often, along the way, those futures pass directly through and dramatically rearrange human lives. There seems to be a growing sense that architectures, infrastructures, and artworks will risk being irrelevant at best, dangerous at worst, if their designers and makers don't take this newly salient reality into account and recalibrate their work.

In Japan, the geologic-as-actant exceeded humans' best attempts to anticipate and temper

Nesjavellir Geothermal Power Station, Iceland, smudge studio 2012

its potential impact on highly technologized daily life. The shattering of these best attempts has opened up new potential for inventing and actually adopting innovative modes of action at the intersections of the human and the geologic. It has necessitated new design specifications for all future, as yet unimagined projects of infrastructure, urban space and flow, and channeling of geologic actants. The fact that two thirds of Japanese citizens now prefer to not have nuclear generated electricity demonstrates that geologic events can instruct humans about the reality that all our designs assemble with and participate in relation to complex, powerful earth forces we can never fully control.

Such "instructive" events seem to be compounding—both actually and within human consciousness—to lay bare the reality of just how deeply human life is embedded in the "brute materiality of the external world"—in the very "stuff" of the geologic. Take, for example: the discovery in 1997 of The Great Pacific Garbage Patch; the Indian Ocean earthquake and tsunami (2004); Hurricane Katrina (2005); Eyjafjallajökull's eruption and subsequent disruption of European air travel and economies (2010); the Haiti Earthquake (2010); Japan's "triple disaster" (2011); the Gulf of Mexico oil spill (2010); Superstorm Sandy (2012); increasing efforts to prepare for the long overdue "big one" along the San Andreas Fault in California; ever-clearer signs of climate change both man-made and earth-made; recent "near misses" of earth by asteroids and a growing understanding of the planet-wide effects of prehistoric direct hits; growing stockpiles of high-level nuclear waste and the urgent attempts to design ways to contain it–within the geologic–for up to one million years; new evidence of how geologic-scale engineering projects, such as the Three Gorges Dam in China and carbon sequestering, actually alter planetary dynamics; and growing evidence that urbanization vastly increases rates of soil sedimentation and erosion. And, as we deliver this collection to our publisher, Australian National University ecologist Roger Bradbury sends this alert to readers of *The New York Times*: "It's past time to tell the truth about the state of the world's coral reefs, the nurseries of tropical coastal fish stocks. They have become zombie ecosystems, neither dead nor truly alive in any functional sense, and on a trajectory to collapse within a human generation. There will be remnants here and there, but the global coral reef ecosystem—with its storehouse of biodiversity and fisheries supporting millions of the world's poor—will cease to be."[3] Far from being yet another instance of "disaster porn," Bradbury sounds his alarm in the name of ending "institutional inertia" and beginning an "enormous reallocation of research government, and

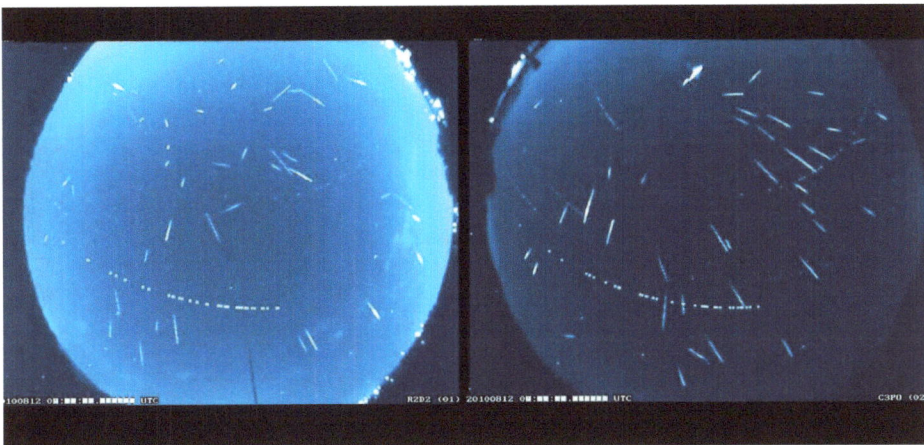

composite of Perseid Meteor shower, 8/11/10, image NASA/MSFC/D, Moser, NASA's Meteoroid Environment Office

---

3  Roger Bradbury, "A World Without Coral Reefs," The New York Times, 13 July 2012, http://www.nytimes.com/2012/07/14/opinion/a-world-without-coral-reefs.html.

environmental effort to understand what has happened so we can respond the next time we face a disaster of this magnitude."

Developments in human thought about the nature of reality itself give additional impetus for the current turn in cultural awareness and practice. These include new scientific understandings of the most fundamental geologic dynamics, such as plate tectonics, which hinges on a theory that some considered impossible as recently as the early 1960s. A turn away from static, mechanical models of "nature" to dynamic models that see the world as a field of continuous emergence has also unleashed both scientific and popular fascination with the deep interplays between the geologic and the biologic. Some philosophers now think experimentally about space, time, and change through the geologic; popular consciousness entertains radical new insights about human evolution, including the realization that life itself is the result of an immensely long process that is fundamentally geologic in nature; and journalists comment on the qualitatively new ways that humans experience time and space on planet earth.

Novel ways of making sense of the geologic have become possible, in part, thanks to new technologies and lived experiences that help to make it immediate. In fact, geology's transformation from being regarded as a specialized area of scientific study at best and nerdy hobby at worst, to being experienced vividly as a daily predicament, may have been ushered in by an event of digital visualization. Arguably, the transformative moment came on December 7, 1972, the day that humans on earth received the transmission of the "blue marble" photograph of earth from the Apollo 17 astronauts across 28,000 miles of space. Thirty years later, the images of artist/photographer Trevor Paglen suggest (as he does in the interview included here) that the earth has gained a new sedimentary layer—located in space—in the form of orbiting space debris.

The internet and globalization have radically transformed daily experiences of time and space on earth in ways that continue to direct attention to the geologic. Philosophers and popular culture producers alike ponder the consequences of global flows and exchanges of information, human beings, manufactured products and earth materials for how humans sense and make meaning of time and space.

Thanks to Google Earth, our abilities to visualize earth as a wholly interconnected and dynamic geo-bio system have grown only more acute, refined, and now even interactively palpable. In addition, dynamic imaging technologies allow us to "see" and "sense" unimaginably slow and vast geologic dynamics as visible "flows." Computer simulations and data animations make apparently static and solid geo-forms such as mountains and deserts perceptible as fluid motion through time.

In the 1980s, word started to spread beyond scientific journals that the geologic conditions of life on earth can be—and have been—transformed in an instant: dinosaurs, some scientists announced, probably went extinct because of an asteroid impact. Other scientists think dinosaurs were wiped out by massive volcanic eruptions in what is now India. Regardless, their disappearance happened fast and most likely because of a geologic event that was planetary in scale. Journalists and Hollywood filmmakers have peaked audiences' curiosities—in ways that add to the scientific knowledge of laypersons—about these very real (past and future) geologic processes. Recent best sellers that help us think across biology, geology, and environmental sciences, such as *The World Without Us*, fuel growing realizations about deep time: while the human species can't get along without the geologic, the geologic will continue on in some form or other long after we have ceased being part of it. Scientific facts about how the geologic is capable of throwing our entire species into a crisis of existential risk are now part of pop culture (the film *2012* offers a special effects version of everything from

geothermal site, Iceland, smudge studio 2012

Noah-sized tsunamis to the movement of tectonic plates to the periodic and believed-to-be-overdue geomagnetic reversal of the earth's poles).

Thanks to transdisciplinary exchanges across traditional fields of study such as science and philosophy, art and science, environmental studies and social sciences, it is becoming difficult for geologists and biologists to hold to categorical distinctions between the "brute materiality" of geology's "external world" (rocks, minerals, mountains) and the soft, "inner" worlds of biology's living things. According to current scientific narratives about life, earth, and life on earth, it's possible to claim, without taking too much poetic license, that we humans are walking rocks. We may be living creatures, but our aliveness is composed of geologic materials such as calcium, iron, and phosphorous. And the comparatively tiny living organisms that inhabit the earth's surface, be they humans, lichen or bacteria, are now seen to be key players in setting up and precipitating monumental geologic processes and planetary-scale chemical transformations in geologic materials. The earth would have a completely different geologic self if there were no life on it.

Such vivid experiences—firsthand, mediated, and imagined—make it increasingly possible, even necessary, for humans to further heighten our abilities to sense and respond to the vastness, and to the agency, of geologic time. Unprecedented modes of inhabiting the earth have opened up further possibilities for humans to evolve ways to live in relation to geologic time.

What we might make of our growing abilities and needs to sense geologic time, and what they might make of us, have become questions charged with more than a little urgency.

These, then, are some of the events, developments, and ideas that actively shape situations and structures of feeling within which contemporary cultural producers are making their work. Contributors to *Making the Geologic Now* have been drawn to the power and promise of a heightened collective awareness of the geologic as a condition of contemporary life. Their works and ideas both signal and make more of the fact that the geologic is a vibrant force in contemporary life.

MAKING THE GEOLOGIC NOW: INTENTIONS, MOTIVATIONS, PROVOCATIONS

Opera House, Oslo, Norway, smudge studio 2011

*We believe that this collection may be the first to offer early sightings of an emergent cultural sensibility: a geologic turn in contemporary ideas, architecture and design, and aesthetic work.* We have designed this book in the tradition of a broadside or pamphlet—a timely "calling out" of an emergent cultural sensibility and practice. Its publication by punctum books as download-able file, bound book, and interactive website makes it readily and widely accessible, portable, and easily exchanged. We hope that these forms will encourage it to move through culture the way the "geologic turn" is now propagating through contemporary consciousness and practice.

As an early sighting of an emergent and expanding cultural sensibility, this collection functions as a provisional and experimental first naming. We do not pretend to provide an authoritative description or systematic mapping of a particular moment in cultural history. Our editorial perspective is not that of a geologist or art critic. This collection constructs a viewpoint that is generative rather than critical or analytical. It is a relay—a strategy to pro-vide impetus to our own and our contributors' enthusiasms and desires for a geologic turn, and to pass them on to others.

Our approach is to present our contributors' ideas and images as test sites—places to think experimentally about what might happen, what might become thinkable and possible, if we humans were to collectively take up the geologic as our instructive co-designer—as our partner in designing thoughts, objects, systems, and experiences? The collection provides an armature for framing responses to such a question. It gathers a broad range of examples of what might count, in thought and practice, as "making the geologic now." As an uneven engagement with an emergent phenomenon, it provides points of departure rather than de-limited categories. Contributors describe their first whiffs of what is barely understood and not yet fully articulated about the fact that the geologic has become palpable as a vibrant force in play in everyday life (once again? in unprecedented ways? more intensely?). After all, any current turn to the geologic in attempts to understand the present moment and map possible

National Tourist Route, Lofoten, Norway, smudge studio 2011

ways forward would be *both* an evolution in human relationship to the geologic, *and* a return to the age-old problem of the place of the geologic in human life on this planet—a problem that the West has both tried to control and now has become. Each essay, then, suggests an optic—an aesthetic and/or conceptual aperture—that readers can use to bring into sight and into mind what is as yet still arriving into human awareness, language, and gesture about the geologic *this* time.

Contributors offer images, designs, projects, practices, and concepts for projecting imaginations into the geologic-as-contemporary-situation. These range from poetic readings of scientific concepts, to theoretical engagements with iconic land art, to visual essays of aesthetic and design provocation. Each work attempts to turn our gaze toward what has been operating in the peripheral vision of Western culture's assumptions about the place of the geologic in human affairs. We believe that, collectively, these essays invite several consequential inversions of perception—from seeing the geologic as matter to seeing it as process; from seeing the stuff of the world we live in as being passive object and inert thing to seeing and sensing it as process and as vibrant matter; from perceiving form as ideal, fixed, or achieved to seeing it as motion; from perceiving humans as the culminating achievement of all of geologic time to seeing ourselves as mammals included within the geologic—as living within what is alien and previous rather than as living within a romanticized nature that "endorses human values" (McKay).

When we invited contributors to plot some points along the contemporary geologic turn, we offered a series of themes for guideposts. These included: the material force of geologic time, designing for and imagining deep time, speculative or aesthetic devices for inciting "evolutionary" cognitive shifts in understanding "the geologic" as a contemporary condition, increasing awareness of the Anthropocene, geologic realities of contemporary land use, new aesthetic forms and styles necessitated by works responding to the geologic, collaborations among artists and geologists/geophysicists, visualizing interplays among multiple, complex geo-bio forces, locating the human scale in deep time and space, altered meanings and sensations of "the geologic" in art and popular culture, making monumentally slow change or movement palpable, practices that propel the geologic turn in contemporary art (such as the journey format, field research, embodied engagement with the geologic), form as motion, motion as form, instability and the earth's materiality, the "intelligence" or "agency" of geologic materials. Taken together, the works we share here touch on all of these themes in one way or another.

The essays in Section One, "Signals from the Edge of an Arriving Epoch," announce and consider scientific, social, and poetic implications of the fact that some geologists have declared the arrival of a new epoch named after us: The Anthropocene (*anthropo*, or man, and *cene*, or new). In 2000, Paul Crutzen proposed that geologists update their official view of geological time to include the Anthropocene. This triggered a still continuing debate within scientific communities as to whether the Anthropocene deserves to be incorporated into the geologic timescale. In 2011, Elizabeth Kolbert's essay in *National Geographic Magazine* (and reprinted here) trumpeted: "Enter the Anthropocene—The Age of Man," and presented Crutzen's concept to a lay readership worldwide. Following Kolbert's essay, and thanks to the work of Valeria Federighi and Etienne Turpin, we offer what we believe to be the first publication of an English translation of an even earlier sighting of this new epoch by Italian geologist Antonio Stoppani. Nearly 150 years ago, he declared: "We are only at the beginning of the new era; still, how deep is man's footprint on earth already? . . . You can already count a series of strata, where you can read the history of human generations, as before you could read in the amassed bottom of the seas the history of ancient faunas." Following on Stoppani, Canadian Poet Don McKay's remarkable enactment of "geopoetry" in his essay: "Ediacaran and Anthropocene: Poetry as a Reader of Deep Time," searches out and expresses some of the

Laura Moriarty, *Breccia*, 2009, Encaustic

profound, sublime, terrifying, and even ethical meanings released by the growing impera-
tive to name the present moment "Anthropocene." The essays by Fox, Gilbert, and Osborne
further clear conceptual and aesthetic pathways for the arrival of the Anthropocene.

We have staged Section Two, "Shifts in the Material Conditions of Contemporary Life," as
a concatenation of gestures by designers, cultural producers and critics, artists, and teachers. Each essay in this section is an indicator of sometimes catastrophic, sometimes barely
perceptible change in the material circumstances of contemporary human beings—change
that carries a geologic charge or imprint. Each emanates from a specific condition or event
that is geologically inflected. These range from direct or indirect experiences of geologically
triggered historic events, to particularly volatile or provocative junctures of the human and
the geologic, to personal encounters with geologic forces and materials and pedagogical
activations of the geologic as learning event. Contributors use speculative essays, land use
interpretation, photographic essays, documentary research, and art criticism to describe
qualitative changes in material conditions of daily life. For example, in "Imagining the
Geologic," Norwegian literary critic Janike Kampevold Larsen takes readers on a journey from
the Grand Canyon to highway road cuts, and offers them a visceral sense of her argument: current
conditions of life are forcing human attention away from pictorial images and views of landscape and toward the bare materiality of the earth's surface. Such a redirection of attention
heightens the possibility that "slight efforts of the imagination" can help us relate to the
ground as a presence that exists both within *and* outside of human mediation. Designer and
researcher Chris Rose offers a first hand account of such an effort, when his own perception
became "inverted" as a result of making a geologic turn: in the midst of conducting field
research with students in a quarry on the coast of England, he found himself newly capable of
experiencing the rock that surrounded him as "process." In "Robert Smithson's Abstract Geology: Revisiting the Premonitory Politics of the Triassic," philosopher and aesthetic theorist
Etienne Turpin sights a precursor to today's broader turn to the geologic as source of provocation to thought and aesthetic practice. He offers the works of Robert Smithson to readers
as points of reference that they can use to help sense their own arrivals into the era of the
Anthropocene.

Essays in this section invite an active reading across diverse ways of responding to geologic
conditions of the present moment, and set up potential for productive collisions among, for

Geopoetry, from *Geologic City: A Field Guide to the GeoArchitecture of New York,*
smudge studio 2011

example, explorations of photography as a particularly effective medium for engaging the force and depth of the geologic, new forms of field expedition and documentary research designed to engage students with the geologic, and speculative or aesthetic responses to pragmatic and existential challenges that humans face as a result of being embedded in the geologic.

Ring Road Bridge, Iceland, smudge studio 2012

In Section Three, "From Periphery to Center: Artists Make the Geologic Now," each work offers a sighting of a place, project, or event in which humans intentionally or unintentionally join with geologic materialities, forms, or processes—and do so in ways that have consequences for both. From monumental sites of extraction, to a forest planted in anticipation of the deep geologic future, to photographic portraits of the new sedimentary layer that all landfills inevitably become, this section indicates how contemporary artists are alerting their audiences to the consequences of human-geologic assemblages.

The fourth section, "Geologic Tomorrow: Wild and Potent Futures," focuses on a particularly fierce geologic material that promises to shape human activities, design challenges, and philosophical thought for the rest of our species' existence. Elizabeth Kolbert's "Enter the Age of Man" appeared in *National Geographic* the same month that Fukushima Daiichi spewed radioactive materials around the globe. This poignant simultaneity gave way to the undeniable fact that the material composition of the planet (ourselves included) is forever altered as of that date. The degree to which such fundamental changes are felt in daily life is not evenly distributed. In mid-2012, we spent a month in Kyoto, Japan and we were surprised by how concerns about radioactive contamination in the air and in water and food supplies in locations hundreds of kilometers south of the accident had become personalized for people living there. We ate often at a Kyoto café that tested their food for radioactive cesium and openly posted the results (see Zuihitsu 2. page 13). Experiencing this first hand and regularly encountering television programs and public events that engaged these concerns, we grasped more fully the degree to which radioactive substances have become a potent geologic materiality—one that is becoming ever-more deeply embedded in individuals' lives. As this section's "Technogeomorphological Mounds" by the Center for Land Use Interpretation, and Geoff Manaugh and Nicola Twilley's interview with a scientist at Yucca Mountain remind us, humans will be "designing for deep time" and attempting to contain landscapes contaminated by nuclear materials in perpetuity. Our species is just at the beginning of a design and engineering project that, if taken up, will span millennia. As Shimpei Takeda, Bryan Wilson, and Jamie Kruse explore, the new stratum containing radioactive materials that we are now laying down always will be traced back to, and always will affect, individual humans.

Today, global flows create complex, moving entanglements of earth materials, geologic events, technologies, objects, chemicals, weather, information, people and other living and nonliving things. Over the past couple of years, we have published a blog (Friends of the Pleistocene) to express and visualize "the geologic" as interwoven with the rise in global populations; nested within the challenges of nuclear waste storage; enfolded in carbon emissions; caught up in the rise of tsunami waves; orbiting the planet as space trash; stuck in the stagnant center of the vortex that is the Great Pacific Garbage Patch; and fermenting in the hills of the Freshkills landfill. *The geologic "now" is a teeming assemblage of exchange and interaction among the bio, geo, cosmo, socio, political, legal, economic, strategic, and imaginary.* The geologic lives in our bones (as calcium) and our cell phone screens (as indium tin oxide). The geologic "now" in which we live, and for which we design urban spaces and infrastructures, is an ongoing procession of substances that were formed in the deep past and are arriving into the present. The geologic passes through our time as the materials and forces that compose us, and that we take up and transform to compose our world. Geo-bio-socio assemblages reconfigure and ramify geologic materials and forces, with growing consequence, into the stuff of deep futures.

back cover, from *Geologic City: A Field Guide to the GeoArchitecture of New York,* smudge studio, 2011

Jane Bennett's work has been an inspiration for the way we have approached the project of *Making the Geologic Now*. Her concluding essay "Earthling, Now and Forever?" responds to the collection in light of key ideas that shape her own work.

Throughout the collection, we have dispersed 15 visual and textual provocations that have affinities with the adjacent essays. We call these interjections "zuihitsu," after a Japanese term that loosely translates as: "a miscellaneous essay," "literary jotting," or "musing." A kanji that is out of common usage for "zui" translates as: "at the mercy of (the waves)." The kanji currently in modern usage in Japan translates as: "follow." Hitsu means "writing or painting brush." Our zuihitsu include original photography and excerpted texts from our own work, our contributors, and various public sources. They provide an unfinished "through line"—a lively and unsettled terrain composed of concepts, images, sensations and waves of realization that "make the geologic now."

Draper, UT, smudge studio 2010, from *Below the Line*

## HUMANS ASSEMBLING WITH THE GEOLOGIC

*We believe that turning toward the geologic, in ways reported and enacted by the essays collected in Making the Geologic Now, has the potential to relocate human sensibilities and assumptions with great effect*. It may, for example, compel activists and designers who invoke the ecological and the "environmental" to take the geologic fully into account. The first wave of environmentalism in the 1970s was dominated by discourses that understood "the environment" primarily in biological terms. It focused its concern on the surface of the earth—on the soil, water, living organisms, and atmosphere that compose the thin layer that is the biosphere. But recent insights about just how deeply the bio is intertwined with the geo make it necessary for any concept of "the environment" to now acknowledge the roles and powers of the geologic in terrestrial life. Given new understandings of the figurative and literal depth of the enmeshment of the bio and the geo, it's now difficult to hold previous distinctions between animal, vegetable and mineral. Current understandings of earth cycles show that the bio is instrumental in the composition, history, and future of the geo. We now know that solar flares, asteroids, and supervolcanoes can alter and have altered the course of biological life on earth in seconds. This awareness extends what we consider to be our environment far beyond Earth's surface. The soci-geo-bio "order" that we live today draws all things on Earth—human and nonhuman—into relation at a much vaster breadth and depth than acknowledged by the environmentalism of the 1970s. Even the landscape or surface that architects design for and build upon can no longer be taken to be that single, thin, folded surface depicted by computer aided design software. Today, the geologic counts as "the environment" and extends it out to the cosmos and down to the Earth's iron core.

A geologic turn may recalibrate nothing less than our senses of temporal and spatial dwelling. It may make us capable, McKay suggests, of an entirely new human relationship to time—one in which we no longer see time only or primarily in relation to humanity's place in it. As we lose our position in/as the pinnacle of all that has been made possible by the material force of geologic time itself, McKay suggests, we may gain new capacities to sense and act in terms of mutuality (McKay).

*By making a geologic turn, we direct sensory, linguistic, and imaginative attention toward the material vitality of the earth itself. We come closer to entertaining the idea that matter is not passive. Assuming its passivity weakens our abilities to discern the force of things* (Bennett, 65). It precludes potentially productive questions such as: how is this geologic actant—this earthquake, geologic stratum, geologic flow, earth materiality—contributing to a problem affecting me? How might these nonhumans contribute to its solutions (Bennett, 103)? We might design and respond differently if we were to work from a place of recognizing that our actions always do assemble with the geologic. This change in perspective may result in wildly unpredictable innovations.

Laura Moriarty, *Breccia (detail)*, 2009

In *Vibrant Matter*, Jane Bennett asserts that it would be advisable, productive, and ethical for contemporary philosophers to "theorize a kind of geological affect or material vitality" that recognizes the "shimmering, potentially violent vitality intrinsic to matter" (61). The Japanese earthquakes and tsunamis of 2011 opened up new potentials for humans to invent and collectively adopt new modes of action at the intersections of the human and the geologic. We can no longer relate to the Earth as brute, static material: rocks, mountains, canyons, continents. Mountains are in constant motion. The stuff of rocks is in continuous transformation. The Earth's crust is a conveyor belt that digests continents and regurgitates new landmasses. Earth has a finite life span constrained by its cosmic environment. New understandings of the power of relatively ephemeral geo-bio-socio assemblages have altered our senses of the place we inhabit. No longer the inert matter outside of ourselves that is there to support us and our buildings, the geologic is a cascade of events. Humans and what we build participate in their unfolding.

Making a geologic turn, we create an opportunity to recalibrate infrastructures, communities, and imaginations to a new scale—the scale of deep time, force, and materiality. This would require us to assemble responsively with the non-human scale of geo-forces in play on this planet. Such a move has the potential to turn Western-encultured humans (once again?) toward what is most real about human life on this planet: we are not simply "surrounded" by the geologic. We do not simply observe it as landscape or panorama. We inhabit the geologic.

Copper Cliff, Sudbury, Ontario, smudge studio 2011

We live within it. This means that humans are always forced to come to terms with earth forces, eventually.

But, as the works collected here show us, it can also mean more than that.

Perhaps the qualitatively new and different cultural sensibility that is signaled by those who are "making the geologic now" is this: *as the very medium of human existence, geologic assemblages are vibrant forces, and they are capable of instructing not only architecture and design practices, but everyday life as well.* ■

# SECTION 1: SIGNALS FROM THE EDGE OF AN ARRIVING EPOCH

Elizabeth Kolbert

# 1. ENTER THE ANTHROPOCENE—AGE OF MAN[1]

The path leads up a hill, across a fast-moving stream, back across the stream, and then past the carcass of a dead sheep. In my view it's raining, but here in the Southern Up-lands of Scotland, I am told, this counts only as a light drizzle, or *smirr*. Just beyond the final switchback, there's a waterfall, half shrouded in mist, and an outcropping of jagged rock. The rock has bands that run vertically, like a layer cake that's been tipped on its side. My guide, Jan Zalasiewicz, a British stratigrapher, points to a wide stripe of gray. "Bad things happened in here," he says.

The stripe was laid down some 440 million years ago, as sediments slowly piled up on the bottom of an ancient ocean. In those days, life was still mostly confined to the water, and it was undergoing a crisis. Between one edge of the three-foot-thick gray band and the other, some 80 percent of marine species died out, most of them the sorts of creatures, like grap-tolites, that no longer exist in any form. The extinction event, known as the end-Ordovician, was one of the five biggest of the last half-billion years. It coincided with extreme changes in climate, in global sea levels, and in ocean chemistry—all caused, perhaps, by a supercontinent drifting over the South Pole.

Stratigraphers like Zalasiewicz are, as a rule, hard to impress. Their job is to piece together Earth's history from clues that can be coaxed out of layers of rock millions of years after the fact. They take the long view—the extremely long view—of events, only the most violent of which are likely to leave behind lasting traces. It's those events that mark the crucial episodes in the planet's 4.5-billion-year story, the turning points that divide it into comprehensible chapters.

from *Geologic City: A Field Guide to the GeoArchitecture of New York,* smudge studio 2011

---

1 Originally published in *National Geographic Magazine,* March 2011: http://ngm.nationalgeographic.com/2011/03/age-of-man/kolbert-text.

So it's disconcerting to learn that many stratigraphers have come to believe we are such an event—that human beings have so altered the planet, just in the last century or two, that we've ushered in a new epoch: the Anthropocene. Standing in the *smirr*, I ask Zalasiewicz what he thinks the dawn of this epoch will look like to the geologists of the distant future, whoever or whatever they may be. Will the transition be moderate, like dozens in the record, or will it show up as a sharp band in which very bad things happened—like the mass extinction at the end of the Ordovician?

That is being determined right now, Zalasiewicz says.

from *Geologic City: A Field Guide to the GeoArchitecture of New York*, smudge studio 2011

"Anthropocene" is a neologism. It was coined by the Dutch chemist Paul Crutzen about a decade ago. One day Crutzen, who shared a Nobel Prize for discovering the effects of ozone-depleting compounds, was sitting at a scientific conference. The conference chairman kept referring to the Holocene, the epoch that began at the end of the last ice age, 11,700 years ago, and that—officially, at least—continues to this day.

"Let's stop it," Crutzen recalls blurting out. "We are no longer in the Holocene; we are in the Anthropocene. Well, it was quiet in the room for a while." When the group took a coffee break, the Anthropocene was the main topic of conversation. Someone suggested that Crutzen copyright the word.

Way back in the 1870s, an Italian geologist named Antonio Stoppani proposed that people had introduced a new era, which he labeled the "anthropozoic." Stoppani's proposal was ignored; other scientists found it unscientific. The Anthropocene, by contrast, struck a chord. Human impacts on the world have become a lot more obvious since Stoppani's day, in part because the size of the population has roughly quadrupled, to nearly seven billion. "The pattern of human population growth in the twentieth century was more bacterial than primate," the biologist E.O. Wilson has written. Wilson calculates that human biomass is already 100 times larger than that of any other large animal species that has ever walked the Earth.

In 2002, when Crutzen wrote up the Anthropocene idea in the journal *Nature*, the concept was immediately picked up by researchers working in a wide range of disciplines. Soon, it began to appear regularly in the scientific press. "Global Analysis of River Systems: From Earth System Controls to Anthropocene Syndromes" ran the title of one paper in 2003. "Soils and sediments in the anthropocene," was the headline of another, published in 2004 in the *Journal of Soils and Sediments*.

At first most of the scientists using the new geological term were not geologists. Zalasiewicz, who is one, found the discussions intriguing. "I noticed that Crutzen's term was appearing in the serious literature, without quotation marks and without a sense of irony," he says. In 2007, Zalasiewicz was serving as chairman of the stratigraphic commission of the Geological Society of London. At a luncheon meeting one day, he decided to ask his fellow stratigraphers what they thought of the Anthropocene. Of the 20 in attendance, 19 thought the concept had merit.

The group agreed to look at it as a formal problem in geology. Would the Anthropocene satisfy the criteria used for naming a new epoch? In geological parlance, epochs are relatively short time spans, though they can extend for tens of millions of years. ("Periods" such as the Ordovician and the Cretaceous last much longer, and "eras" like the Mesozoic longer still.) The boundaries between epochs are defined by changes preserved in sedimentary rocks—the emergence of one type of commonly-fossilized organism, say, or the disappearance of another.

The rock record of the present doesn't exist yet, of course. So the question was: when it does, will human impacts show up as "stratigraphically significant?" The answer, Zalasiewicz's group decided, is "yes"—though not necessarily for the reasons you'd expect.

Probably the most obvious way humans are altering the planet is by building cities, which are essentially vast stretches of manmade materials—steel, glass, concrete, brick and cement.

Fresh Kills, from *Geologic City: A Field Guide to the GeoArchitecture of New York*, smudge studio 2011

But it turns out most cities are not good candidates for long-term preservation, for the simple reason that they are built on land, and on land the forces of erosion tend to win out over those of sedimentation. From a geological perspective, the most plainly visible human effects on the landscape "may in some ways be the most transient," Zalasiewicz has observed.

Humans have also transformed the world through farming; something like 38 percent of the planet's land surface is now devoted to agriculture. Here again, some of the effects that seem most significant today will leave behind only subtle traces.

Fertilizer factories, for example, now extract more nitrogen from the air than all natural processes combined; the runoff from fertilized fields is triggering life-throttling blooms of algae at river mouths all over the world. But this global perturbation of the nitrogen cycle will be hard to detect, because synthesized nitrogen is just like its natural equivalent, and algal blooms occur naturally too. Future geologists are more likely to grasp the scale of 21st-century industrial agriculture from the pollen record—from the monochrome patches of corn, wheat, and soy pollen that will have replaced the varied record left behind by rainforests or prairies.

The leveling of the world's forests will send at least two coded signals to future stratigraphers. Deciphering the first may be tricky. Massive amounts of mud eroding off denuded land are deepening ocean sediments in some parts of the world—but at the same time the dams we've built on most of the world's major rivers are holding back sediment that would otherwise be washed to the sea. The second signal of deforestation should come through clearer. Loss of forest habitat is a major cause of extinctions, which are now happening at a rate hundreds to thousands of times higher than during most of the last half a billion years. If current trends continue, the rate may soon be tens of thousands of times higher.

Probably the most significant change, from a geological perspective, is one that's invisible to us—the change in the composition of the atmosphere. Carbon dioxide emissions are colorless, odorless, and in an immediate sense harmless. But their warming effects could easily push global temperatures to levels that have not been seen for millions of years. Some plants and animals are already shifting their ranges toward the poles, and those shifts will leave traces in the fossil record. Some species will not survive the warming at all. Even if most of the Antarctic ice sheet survives, sea level would still rise by fifty feet or more.

Long after our cars, cities, and factories have turned to dust, the consequences of burning billions of tons worth of coal and oil are likely to be clearly discernible. As carbon dioxide is warming the planet, it is also seeping into the oceans and acidifying them. Some time this century they may become acidified to the point that corals can no longer construct reefs, which would register in the geological record as a "reef gap." Reef gaps mark each of the last five major mass extinctions. The most recent one took place 65 million years ago at the end of the Cretaceous Period; it eliminated not just the dinosaurs, but also the plesiosaurs, pterosaurs, and ammonites. The scale of what's happening now to the oceans is, by many accounts, unmatched since then. Future geologists are not likely to mistake the beginning of the Anthropocene.

Jökulsárlón glacier lagoon, Iceland, smudge studio 2012

But when, exactly, did the Anthropocene begin? When did human impacts rise to the level of geological significance?

William Ruddiman, a paleoclimatologist at the University of Virginia, has proposed that the invention of agriculture 8,000 years ago, and the deforestation that resulted, led to an increase in atmospheric $CO_2$ just large enough to stave off what otherwise would have been the start of a new ice age; in his view humans have been the dominant force on the planet practically since the start of the Holocene. Crutzen has suggested that the Anthropocene began in the late 18th century, when, ice cores show, carbon dioxide levels began what has since proved to be an uninterrupted rise. Other scientists put the beginning of the new epoch in the middle of the 20th century, when global population and consumption growth suddenly began to accelerate.

Finally, some argue that we've not yet reached the start of the Anthropocene—not because we haven't had a significant impact on the planet, but because the next several decades are likely to prove even more stratigraphically significant than the last few centuries. "Do we decide the Anthropocene's here, or do we wait 20 years, and things will be even worse?" explains Mark Williams, a geologist and colleague of Zalasiewicz's at the University of Leicester in England.

Zalasiewicz now heads a subcommittee of the International Commission on Stratigraphy (ICS) tasked with making an official determination on whether the Anthropocene deserves to be incorporated in the geological time scale. A final decision on the matter will require votes both by the ICS and its parent organization, the International Union of Geological Sciences. The process is likely to take years.

Crutzen, who started the debate, thinks its real value won't lie in revising the charts in geology textbooks. He wants it to focus our attention on the consequences of our collective actions—on their scale and permanence. "What I hope," he says, "is that the term 'Anthropocene' will be a warning to the world." ■

# ANTHROPOCENE

## APPROX. 1945 A.D. – PRESENT

A GEOLOGIC EPOCH WITH NO PRECISE START DATE.
SIGNIFICANT HUMAN IMPACT ON CLIMATE AND ECOSYSTEMS. COINED BY PAUL CRUTZEN.
RISE OF AGRICULTURE. DEFORESTATION. CEMENT. COMBUSTION OF FOSSIL FUELS.
COAL, OIL AND GAS ROUSED FROM THE EARTH. EXTRACTIONS AND EMISSIONS.
OPERATION CROSSROADS VAPORIZES BIKINI ATOLL. DEEP GEOLOGIC REPOSITORIES.
PACIFIC TRASH VORTEX, A SWIRLING GYRE OF MARINE LITTER AND PLASTIC.
6.7 BILLION HUMANS +.  PALO VERDE NUCLEAR POWER PLANT. THREE GORGES DAM.
FRESH KILLS LANDFILL. LAS VEGAS. DUBAI.

from *Geologic Time Viewer*, smudge studio 2010-12

Edited by Etienne Turpin + Valeria Federighi
Images Lisa Hirmer

## 2. A NEW ELEMENT, A NEW FORCE, A NEW INPUT: ANTONIO STOPPANI'S ANTHROPOZOIC[1]

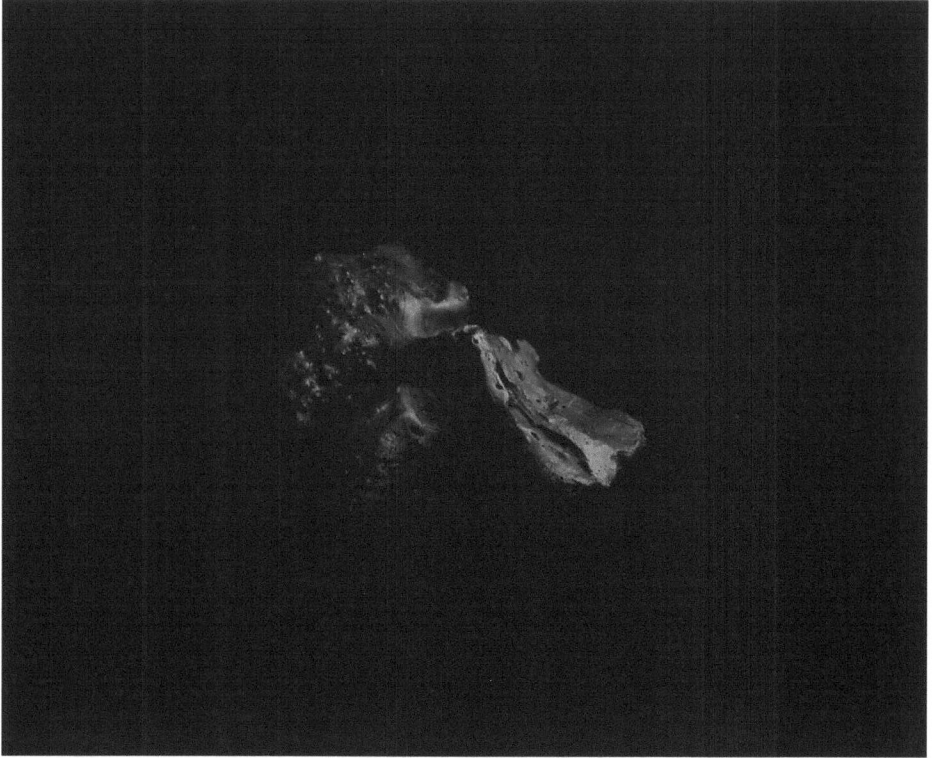

Lisa Hirmer, *Untitled, from Sudbury Slag*, 2012

## INTRODUCTION

The Italian geologist Antonio Stoppani is a remarkable but little known figure in the history of science and the theoretical humanities. Recently, following debates about the Anthropocene initiated by the Dutch chemist Paul Crutzen, some scholars have returned to Stoppani's writing for its eloquent argument regarding the appearance of human activity in the archive of deep time—the earth. Born in Lecco in 1824, the young Stoppani studied to become a priest of the order of the Rosminiani, and was ordained in 1848. In the same year, Stoppani participated in the resistance during the Cinque giornate di Milano (Siege of Milan), where he both fought on the barricades and, fantastically, invented and fabricated aerostats that were used to communicate with the periphery and the provinces, sending revolutionary messages to the countryside from inside a barricaded Milano. In this endeavor, he was helped by the typographer Vincenzo Guglielmini, who worked with Stoppani to ensure that the aerostat balloons would travel from the Seminario Maggiore di Porta Orientale over the walls erected around the city (and the Austrians trying to shoot them from the sky) to encourage Italians to revolt against the Austrian Empire.

---

1 Excerpts from Antonio Stoppani, *Corso di Geologia*, trans. Valeria Federighi, ed. Valeria Federighi and Etienne Turpin (Miliano: G. Bernardoni, E G. Brigola, Editori, 1873). Photographs by Lisa Hirmer.

Following this siege, Stoppani also participated in subsequent confrontations, but after the Battle of Novara he returned to the seminary as a grammar teacher. This return was short-lived, however, because Stoppani's patriotic past and political ideas remained unwelcome by the Church. Following his expulsion from the seminary, he began to study geology. And, while his religious conviction is clear and consistent in his writings on geology, it is for his advances in understanding terrestrial affairs, not theological dogma, that he is best remembered. Notably, after the liberation of Milan, Stoppani's merits were acknowledged and his old titles reinstated. In 1867, he was appointed Professor of Geology at the Politecnico di Milano, where he also helped to found the Museum of Geology, and acted as president of the Geological Society. An experienced alpinist, in 1874 Stoppani became the first president of the Milan section of CAI (Club Alpino Italiano).

In the late 1880s, Stoppani would return to confront his theological roots, publishing *Gli intransigent*—a book critical of the Catholic Church and its resistance to political and social change—which prompted attacks from L'Osservatore Romano. Later, in his ethnographic study of the various places and populations that inhabited the recently unified Italian territory, *Il bel paese*, Stoppani would wonder at the diversity of tellurian physical expression: "Italy is almost—I don't stammer in saying this—the synthesis of the physical world." The excerpt below, translated from Stoppani's three-volume *Corso di Geologia* of 1873, is an example of his breadth of knowledge, courageous imagination, and compelling but accessible rhetorical inventiveness. Nearly thirteen decades before Crutzen's coinage of the Anthropocene, in this text we find an untimely assessment of the human relation to deep time; perhaps, in the wake of these more recent debates, we finally have ears to hear him.

Lisa Hirmer, *Untitled*, from *Sudbury Slag*, 2012

## FIRST PERIOD OF THE ANTHROPOZOIC ERA

by Antonio Stoppani, translated by Valeria Federeighi,
edited by Etienne Turpin + Valeria Federeghi
Images Lisa Hirmer

I recall with pleasure the event that we believe opened the vulgar era. When was it (more for a necessity as felt by the universe, than for a convention accepted by historians of all nations) that we began to count years anew, and we established the two eras, in which we partition universal history? This happened when in the world resounded the great Word; when, in the bosom of the aged fabric of ancient pagan societies, the Christian ferment was introduced, the new element *par excellence*, that substituted ancient slavery with freedom, darkness with light, fall and degeneration with rebirth and true progress of humanity.

It is in this sense, precisely, that I do not hesitate in proclaiming the Anthropozoic era. The creation of man constitutes the introduction into nature of a new element with a strength by no means known to ancient worlds. And, mind this, that I am talking about physical worlds, since geology is the history of the planet and not, indeed, of intellect and morality. But the new being installed on the old planet, the new being that not only, like the ancient inhabitants of the globe, unites the inorganic and the organic world, but with a new and quite mysterious marriage unites physical nature to intellectual principle; this creature, absolutely new in itself, is, to the physical world, a new element, a new telluric force that for its strength and universality does not pale in the face of the greatest forces of the globe.

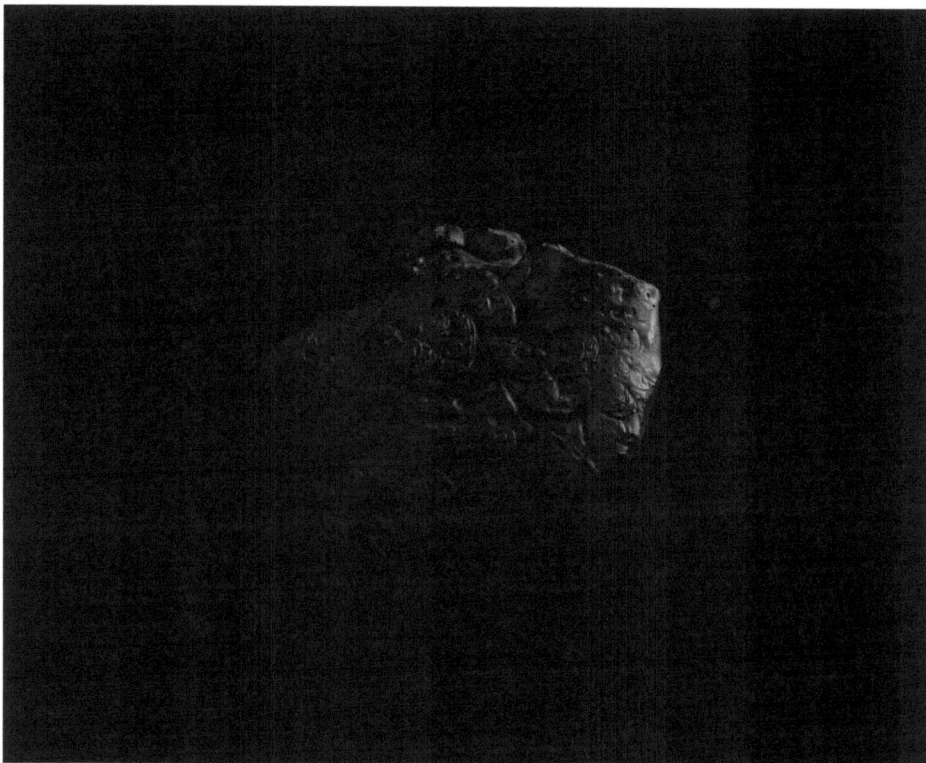

Lisa Hirmer, *Untitled*, from *Sudbury Slag*, 2012

Geology, too, feels thrust onto a new path, feels that its most powerful means, its surest criteria, fail: it becomes, too, a new science. Already the Neozoic era forced it to walk very differently than how it had walked when it only narrated the most ancient events. The science of ancient seas was already destined to become the science of new continents. But even this road cannot lead geology to its destination. It is not enough to consider earth under the impetus of telluric forces anymore: a new force reigns here; ancient nature distorts itself, almost flees under the heel of this new nature. We are only at the beginning of the new era; still, how deep is man's footprint on earth already! Man has been in possession of it for only a short time; yet, how many geological phenomena may we inquire regarding their causes not in telluric agents, atmosphere, waters, animals, but instead in man's intellect, in his intruding and powerful will. How many events already bear the trace of this absolute dominion that man received from God when, still innocent, first heard those words: *Be fruitful and multiply, fill up the earth and subdue it; and rule over the fish of the sea, the bird of the sky and every living thing that moves on the earth, and when, guilty, he heard said: You will earn your bread with your sweat?*

To understand how deep the changes brought about on the globe by this new element are, and how new, consequently, the criteria that guide science should be, it should suffice to make a comparison between so called virgin lands (if there are still any that deserve that name) and those that have been cultivated for centuries. Let us look at Europe, where man has pushed his dominion most forward and where, although recent, his footprints are the deepest.

If his power could do nothing against the strength of the winds, which lead seawaters into the fields that he farms, nonetheless he extends his dominion over the waters themselves as soon as they sprout from the cumuli that wonder in the atmosphere. From the humble brook, that springs from cliff to cliff, to the river that widens its mouth as it debouches into the sea, all flowing waters, oblivious of ancient laws, beat the path that man has traced for them. The old alluvial expanses, already beaten by them with whirling winding, and drowned by their overflowing floods, subtracted by force to their capricious domain, are converted into

Lisa Hirmer, *Untitled*, from *Sudbury Slag*, 2012

greening meadows and fertile fields, periodically mowed by their new owner. Where natural valleys truncate, artificial valleys begin that man traced, guiding gigantic banks along lines as long as are those dug by the slow labor of centuries; and if a river, in the end, finds anew the bosom of the ancient sea, it will be through a different mouth. Waters are not safe, even when they flow furtive underground. Man chases them, catches them, then fountains and rivers, to which man imposes the name of wells, quench the flock's thirst and irrigate the desert. At the same time he severs springs to the exuberant superficial waters, and disperses them into his cisterns.

Already there are new mountains where old valleys used to be: already the irregular soil is drawn into wide plains where waters extend into a thin veil. Already the impenetrable Alps have heard the chisel and the mine resonate in their bosom, and nations have kept a lookout in order to brotherly shake hands. Everywhere, the bosom of the ancient Mother discloses, and the shadows, broken by vagrant splendors, resign to man treasures that were hidden by centuries. At times you can see this Prometheus awaken fire from the bowels of the earth, and guide it to his furnace. Rival of the potent agents of the internal world, man undoes what nature has done. Nature has worked for centuries at agglomerating in the bowels of the earth oxides and metallic salts; and man, tearing them out of the earth, reduces them to native metals in the heat of his furnaces. In vain you would look for a single atom of native iron in the earth: already its surface is enclosed, one could say, within a web of iron, while iron cities are born from man's yards and float on the sea. How much of the earth's surface by now disappears under the masses that man built as his abode, his pleasure and his defense, on plains, on hills, on the seashores and lakeshores, as on the highest peaks! By now the ancient earth disappears under the relics of man or of his industry. You can already count a series of strata, where you can read the history of human generations, as before you could read in the amassed bottom of the seas the history of ancient faunas. To the archeolithic strata, where human relics appear as buried among cut firestones and bones of disappeared animals, terramare superimpose, and pile dwellings, this is where the progress of human race is testified by bronze forged into exquisite weapons and tools. Yet we have not come to see the soil imprinted upon by Etruscan art; and to find ourselves on our own, we have to cross the immense stratum that carries the mark of Roman genius. The rivers, almost oblivious to old granite and porphyry pebbles, learned how to roll pottery and crockery. In the end, approximately 300 million are the men that work, bent and sweaty, from morning until night, on the soil of this small parch of the earth's surface that is called Europe. England, where human industry is the most fervent, crumbles and caves in, everywhere eaten through by insatiable coal, rock salt, limestone and metal miners. What will happen, when Europe will all be worked through as England, and the whole world as Europe? Furthermore, man's influence is not limited to dry land. The very sea cannot escape his dominion. It recedes already, pushed back by obtrusive dams, and by pumps, and by joints, that steal from it arms and lagoons and swamps to make fields. Neither is its immensity of any help in dividing land from islands, islands from continents, as thousands and thousands of ships have opened the way through which nations can embrace, and lands exchange products of the three kingdoms in mutual tribute. Even the unexplored depths of the ocean were forced to act as intercessor, in order to put in contact the peoples of the two worlds. And man invades the atmosphere as well, and not content to only pour, as animals do, the products of his respiration into it, he also pours vast amounts of the products of his industry, gases from his fires and his grandiose laboratories. A century, or just a year, since a family of men settles onto virgin soil, and everything is changed, everything breathes with the strength of human intelligence.

So man dominates over inorganic matter and over forces that alone had governed him for innumerable centuries; but his yoke does not spare the other, nobler, kingdoms. The iron law that his sin brought upon him made man essentially, though other diverse names, a farmer. Here he razes woods; there he covers bare lands with woods; wood is turned into tools; logs into poles; deserts become meadows; squalid moors, verdant fields; nude hills, vineyards and gardens. Greens are not allowed to grow haphazardly any longer, nor to agglomerate into messy and nameless groups. Arranged in rows, seeded in beds, grouped in woods that take their names from the essence that man planted there, cut, pruned, tormented in innumerable guises, fed by artificial heats and waters, they testify everywhere that man has taken full control of that kingdom which God has allocated him for food and shelter. Neither, under his irresistible strength, have plants only submitted to a regime that nature had not imposed; but, oblivious to their own primitive nature, bowing to forced matrimony, new species are simulated under the horrific mask of hybridism while others lie with the flowers and fruits that grafting created. Botanists can only look into the furthest depths, into mountains' fissures, on the highest peaks, for the untamed daughters of virgin nature, that carry unaltered the features of their mother.

Lisa Hirmer, *Untitled*, from *Sudbury Slag*, 2012

We are talking about European man, because Europe, more than other regions, feels man's sovereignty. Home to ancient civilizations, occupied by powerful nations, by men used to multiply time through the zeal of labor, it felt more than others the deep footprint of the earth's lord. But the ancient civilizations of Asia and Africa preceded the ancient civilizations of Europe. The civilizations of Peru and Mexico get lost in the mist of time. Europe, as by regurgitations, flows now over the lands from which its own people originated, and over those, which our fathers didn't know the existence of. For a long time now this wave of people goes, comes, returns, bumps, overlaps as sea waves on the surface of land. Let us not forget, then, that man has been, since his incipit, cosmopolitan. Unlike speechless animals that preceded him on the surface of the planet, he knows no geographical confine, he makes no distinction of zone or of climate; rivers, seas, valleys and mountain crests are no obstacle to him. As he has been wandering for centuries, naked, through the arenas of the boundless desert; so, covered in skins torn from mild and ferocious animals, for centuries he has been driving his sled on the horrid labyrinth of polar ices that reflect the meek glow of the northern lights. But European man already cast his eye on the heart of the desert to make an oasis for himself, and is about to drive his banner on the North Pole, that same banner that already waves on the highest Alpine peaks. A day will come, when the earth will be but a seal of man's power, and man a seal of God's, who, giving man his own image, almost gave him a portion of his own creative will.

A new era has then begun with man. Let us admit, though eccentric it might be, the supposition that a strange intelligence should come to study the Earth in a day when human progeny, such as populated ancient worlds, has disappeared completely. Could he study our epoch's geology on the basis of which the splendid edifice of gone worlds' science was built? Could he, from the pattern of floods, from the distribution of animals and plants, from the traces left by the free forces of nature, deduct the true, natural conditions of the world? Maybe he could; but always and only by putting in all his calculations this new element, human spirit. At this condition, as we, for instance, explain the mounds of terrestrial animals' bones in the deep of the sea, he, too, could explain the mounds of sea shells that savage prehistoric men built on the coasts that they inhabited. But if current geology, to understand finished epochs, has to study nature irrespective of man, future geology, to understand our own epoch, should study man irrespective of nature. So that future geologist, wishing to study our epoch's geology, would end up narrating the history of human intelligence. That is why I believe the epoch of man should be given dignity of a separate new era.

Lisa Hirmer, *Untitled, from Sudbury Slag*, 2012

Geologists should not be reluctant in accepting this foundation for the only reason of the brevity of time currently encompassed by it. The Anthropozoic era has begun: geologists cannot predict its end at all. When we say Anthropozoic, we do not look to the handful of centuries that have been, but to those that will be. Nothing makes us suspect that Adam's seed might be close to extinguishing; for humanity is too young if compared to that ideal of perfect civilization of which mankind's first-born has planted the seed, surely not in vain. But as contained as the number of centuries God is willing to concede to the triumph of intelligence and love may be, the earth will never escape the hands of man if not thoroughly and deeply carved by his prints. The first trace of man marks the beginning of the Anthropozoic era. ∎

Lisa Hirmer, *Untitled*, from *Sudbury Slag*, 2012

William L. Fox
Images William Lamson

# 3. FROM ROCK ART TO LAND ART / FROM PLEISTOCENE TO ANTHROPOCENE

Image courtesy of William Lamson, *A Line Describing the Sun*, 2010

Our species of primate, *Homo Sapiens*, has been around for perhaps 200,000 years, and since then humans have witnessed two geologic turns. The Pleistocene Epoch lasted from approximately 2,588,000 years to 12,000 years before the present (BP), a period of repeated glaciation during which ice at maximal times covered as much as 30% of the Earth's landmasses. The Pleistocene turned into the Holocene as the seemingly perpetual El Niño conditions switched off and the ice retreated. The "Recent Time" of this next epoch saw the spread of humanity into virtually every corner of every continent, save the Antarctic. The second turn we've seen is from the Holocene into the Anthropocene, which Nobel Laureate Paul Crutzen proposes began with the late 18th-century rise of industrialism. This "Human Time" is marked by a global layer of carbon laid down by the burning of fossil fuels, the emergence of a global stratigraphic event being a commonly accepted marker of a change in epochs.

When Paul Crutzen announced in 2000 that we were no longer living in the Holocene, but in a new geologic epoch, he reformulated not only the equation of anthropic effect on landscape, but the timeline of art as we understand it. Now the Holocene was not simply the epoch of human endeavor and image making, but a bridge from the hominids who scratched marks on rock surfaces to those who now leave larger marks inscribed on the land. Both can be construed as attempts to map environments, be they at the small scale of a valley or the large scale of a cosmos, committed with intent to last for millennia or to wash away in the first rain.

Australia is the oldest continent, some of its rocks dating back to 3.5 billion years, and the date of its occupation by humans is steadily pushed back by anthropologists almost every other year. Human arrival on the continent was estimated at 40,000 years BP during the 1990s, but some scientists are now proposing a date closer to 60,000 years BP. Some of the rock art of Australia pictures animals extinct as long ago as 40,000 years BP (for comparison, the earliest animals painted in the Chauvet cave of Southern France, which are the oldest known cave paintings, may date back to between 30,000 and 32,000 years BP). The Dampier Archipelago of Western Australia is the location of the one of oldest and perhaps the largest rock art sites in the world with as many as a million petroglyphs densely inscribed on its islands and the Burrup Peninsula. At the foot of the peninsula is a fenced enclosure behind which rest several hundred red rocks bearing petroglyphs, placed there in the early 1980s in an attempt to preserve the rock art from the development of an immense liquid natural gas processing facility. The 1800 or so boulders are mostly upside down to preserve the images from corrosive gases and natural weathering.

It's shocking to see Pleistocene art that's been torn from its site—thus its cosmological significance to Aboriginal people—then turned away from the sky and imprisoned behind a chainlink fence. But the human activity that creates the Anthropocene, in particular resource extraction, has a way of doing that, of turning things upside down. The Anthropocene is defined as an epoch during which humans literally brought forward out of time energy locked up in old rocks in order to expend it here and now. The most recent geologic turn is marked literally with turning things upside down and inside out.

The most renowned and cited geoscientist of the 19th century was Alexander von Humboldt, whose 5-year excursion to South America with the artist-botanist Aimé Bonpland midwifed the birth of Earth systems science and ecology (and an early proposal for continental drift). But his magnificently illustrated accounts of the journey, published in thirty volumes with the help of fifty some illustrators, engravers and calligraphers, also inspired several generations of artists to paint landscapes that were not only brilliant depictions of geological wonders, but encyclopedic depictions of eco-regions. Frederick Church's canvases of the Andes and Thomas Moran's of Yellowstone and the Grand Canyon are notable examples. Von Humboldt's scientific project, which encompassed or outright invented entire fields from physical geography to meteorology, were anchored upon his training as a mining engineer, which is to say geology.

Image courtesy of William Lamson, *A Line Describing the Sun*, 2010

As anthropic effects upon landscape became more noticeable through the artifacts of the Industrial Age being strewn about the landscape, documentary artists took note. Increasingly those artists were photographers, from Timothy O'Sullivan to the New Topographics artists of the 1960s, such as Robert Adams and Hilla and Berndt Becher. Now the typologies being established were of built structures, such as tract housing and blast furnaces, highway overpasses and retaining walls. What von Humboldt and his colleagues had established in the 19th century was the overarching idea that to picture the world was to necessarily image a system of life intricately and inescapably linked to the geologic. The idea of geographical determinism—that who you are and what you build are profoundly influenced by where you live—is extended in contemporary terms by scientists as notable as E.O. Wilson and geographer Jared Diamond. And that was as true of the rock art makers of the Burrup Peninsula as it would be for the artists of Earthworks in the 1970s.

While Ansel Adams was carefully manipulating his photographs at mid-century to excise evidence of human habitation in the landscape, and the New Topographers were making views of suburban sprawl, another shift was occurring among artists reacting to the Anthropocene, and that was to use land itself as a medium. Earthworks artists such as Michael Heizer and Robert Smithson were aware of both geology and the changes being wrought upon it by humans. Smithson's romanticism of geology was constant, leading him to define the context for his most famous work, *Spiral Jetty*, as "fluvial entropy," an acknowledgement that the geomorphology of the Earth was constantly in change. Pushing his curved line of boulders, an obvious appropriation of a rock art symbol, out from the shore next to an abandoned oil drilling site was more than just an ironic gesture: it was an acknowledgment of how geological epochs were linked.

Heizer's father, the renowned archeologist Robert F. Heizer, was an expert on everything from the stone structures of the Olmec people in southern Mexico to the rock art of the Great Basin, and the sculptor's family included mining engineers, as well. The human history of moving large stones is exactly why Heizer's newest work at the Los Angeles County Museum of Art involves shifting the most massive geological object ever moved by humans, a 350-ton, 21' 6"-tall granite boulder transported on a truck with 200 wheels from a Riverside quarry to Wilshire Boulevard where it will become *Levitated / Slot Mass*. It's another example of an artist of the Anthropocene unearthing and reorienting a piece of geology. People will walk under the 680,000-pound boulder through a trench 456 feet long and 15 feet deep, putting themselves under geology, Heizer using sheer size to recalibrate their sense of scale in the world.

Heizer's largest project, *City*, in Nevada's south-central Garden Valley, is more than a mile long and consists of a wide trench 20 feet deep. Your isolation from the landscape, even as you are immersed in it, is enhanced by berms 20 feet high, the trench and its walls enclosing concrete forms of both Mesoamerican and European origin in a stringent dialogue between the New and Old Worlds. But to climb up the face of one of the huge concrete-shot

Image courtesy of William Lamson, *A Line Describing the Sun*, 2010

*stelae* forms from the bottom of the sculpture to its top is to rise straight out of the valley sediments of the Pleistocene into the fumes of the Anthropocene. Michael Heizer's body of work, which includes numerous sculptural enlargements of tools from the late Pleistocene excavated by his father, is a complete bridge across the Holocene, but it's not the only example.

Land Arts has evolved away from large static projects, such as those by Heizer and Smithson. A number of them, for example, have been participatory projects based on dynamic celestial interactions, including Nancy Holt's *Sun Tunnels*, James Turrel's *Roden Crater*, and Lita Albuquerque's *Stellar Axis* in the Antarctic. These, too, have roots in the rock art of the Pleistocene, many of them, including spiral petroglyphs in the American Southwest, being the target for shadows cast on equinoxes and solstices by carefully placed stone uprights and natural features.

A recent project that fuses the Pleistocene to the Anthropocene is William Lamson's *A Line Describing the Sun*. Using a large Fresnel lens mounted on a wheeled conveyance, Lamson pushed his instrument across the Harper Lake playa in the Mojave Desert and melted a line across the polygon-patterned ground with a 1600°F ray of light. Lamson used real-time sunlight, versus that stored in fossil fuel, to burn a thin black arc 366 feet long into the Late Pleistocene sediments. The daylong performance, captured and condensed into a 13:35 minute video, has another layer of meaning in that it was conducted on the same dry lakebed that houses part of the world's largest solar energy plant. NextEra Energy Resources array of 900,000 mirrors spread out over 1500 acres was built specifically in response to market conditions dictated by the rising energy costs and pollution of the Anthropocene. Lamson, instead of using sunlight to picture the world in a photograph, puts the energy to work directly on the land-as-medium.

The art of the Anthropocene can be seen as a reaction executed by humans to how we perturb earth systems at the geological level. It's not that Heizer, by re-situating a 340-ton boulder over our heads, is deliberately pointing that out in his work, but it has that effect. It allows us to conjure up the connections from one artistic epoch to another. Von Humboldt, who studied ancient ethnographic arts as well as discussed modern color theory with Goethe, would have had no trouble understanding how humans use geology—from boulders as a canvas to a photograph of boulders to a boulder levitated into the heavens—to constantly renew our picture of the world. ■

Image courtesy of William Lamson, *A Line Describing the Sun*, 2010

Don McKay

## 4. EDIACARAN AND ANTHROPOCENE: POETRY AS A READER OF DEEP TIME[1]

Photo: Phoebe Cohen

Two new developments in the taxonomy of temporality provide the focus for this discussion. One is the official recognition of a new, and very old, geologic period, the Ediacaran, now understood to occupy the stretch of deep time between 575 and 542 million years ago, directly preceding the Cambrian, with its remarkable radiation of life forms. This recognition is precipitated by the discovery, dating, and analysis of thirty or so species representing an entirely new biota in the fossil record, the earliest animals on the planet. The other wrinkle in our idea of time involves the proposal to name, or re-name, the current epoch after the species that has been most responsible for its character and style, as well as the content of most of its narratives. "The Anthropocene," if accepted, would acknowledge ourselves as the superstars we have been for some time.

After its artful colon, the title divulges the unusual approach that I propose to take. My inspiration for this notion—reading elements in deep time poetically—comes from an unlikely source. It is in fact a geologist, Harry Hess, who coins the handy term "geopoetry," a term that will certainly serve to identify the path I'm attempting to follow here. Hess was one of the researchers whose work led to the breakthrough understanding of plate tectonics, the crucial concept of a dynamic planet that revolutionized earth sciences in the 1960s. He described his speculations as geopoetry in order to induce his readers (mostly other geologists) to suspend their disbelief long enough for his observations about seafloor spreading, driven by magma rising continuously from the mantle, to catch on. He needed his audience, in the absence of much hard data, to speculate imaginatively, as if reading poetry. Now that

---

1  Reprinted from *Prairie Fire* 29.4 (Winter 2008-2009): 4-15 [The Anne Szumigalski Memorial Lecture].

so much evidence is in, and no one disbelieves in plate tectonics any more (at least no one who does not also disbelieve in evolution), the term might be allowed to lapse, a marriage of convenience whose *raison d'être* has evaporated. But, as you can see, I don't think it ought to be. I think that Harry Hess, like Charles Darwin, Albert Einstein, or any other creative scientist, enters a mental space beyond ordinary analysis, where conjecture and imaginative play are needed and legitimate, and that this is a mental space shared with poets. But even more than this poetic license, I would say, the practice of geopoetry promotes astonishment as part of the acceptable perceptual frame. Geopoetry makes it legitimate for the natural historian or scientist to speculate and gawk, and equally legitimate for the poet to benefit from close observation, and from some of the amazing facts that science turns up. It provides a crossing point, a bridge over the infamous gulf separating scientific from poetic frames of mind, a gulf which has not served us well, nor the planet we inhabit with so little reverence or grace. Geopoetry, I am tempted to say, is the place where materialism and mysticism, those ancient enemies, finally come together, have a conversation in which each hearkens to the other, then go out for a drink. This may not lead to marriage or even cohabitation, but I'm guessing it does lead to a series of dates, trysts, rendezvous, and other encounters whose mood is erotic rather than simply disputatious.

First, the Ediacaran period. The first new period to be introduced to the geologic time scale in 120 years, it is, as Dr. Guy Narbonne has said, equal in importance to the discovery of a new planet in the solar system. The International Union of Geological Sciences has taken this measure because the fossils from the period, pre-dating those of the Burgess Shale, open an entirely unread chapter in the history of life. Ediacaran sites are rare, since these animals (or, some suggest, these members of an entirely new kingdom as yet unknown to taxonomy) were soft-bodied creatures without the shells and hard body armour, which make arthropods like trilobites common in the fossil record. As is usual in the naming of geologic periods, the name derives from a site where the index fossils or strata are found. So the Jurassic period derives its name from the Jura region in Switzerland, the Permian from the city of Perm in Russia, the Cambrian from Cambria, or Wales. Ediacara is in Southern Australia, where these fossils were first discovered, but some of the world's best examples are at Mistaken Point, on the southern tip of the Avalon Peninsula in Newfoundland. In fact there are a few examples of the species known as *Aspidella terranovica* in downtown St. John's on an outcrop kitty-corner from Tim Hortons. As an aside, I wish to observe, as a would-be geopoet, that it is too bad the Mistaken Point fossils were not discovered, or recognized for what they were, before those in Australia, since it would be a splendid moment in the annals of taxonomies to have a period called the Mistaken Pointarean. The name would carry an implicit awareness of its own instability, a fine thing in a name, if you ask me. Mistaken Pointarean would also be appropriate, perhaps, because these creatures seem to have survived a mere 50 million years, an eye-blink in deep time, and only something like 49 and three quarters million years longer than our own distinguished genus. I intend to circulate a petition asking the International Union of Geological Sciences to make this change in the interests of poeticizing the nomenclature. At the very least, the tourist board of Newfoundland and Labrador should support the move, aiming to reap some of the tourist dollars no doubt enjoyed by the likes of Jura and Perm.

If you were to travel as a geopoet-in-training to Mistaken Point to see these remarkable fossils, you would also see the beautiful barrens of the South Avalon on your hike into the site. These barrens—themselves a candidate for renaming by poetry—are only bare if your idea of flora excludes everything under four feet high. The carpet of vegetation over these windswept heaths is an interwoven mat of crowberry, partridgeberry, cranberry, juniper, sheep laurel, bottlebrush and Labrador Tea, with sporadic groves of tuckamore dotted here and there.

Horned Larks and Water Pipits materialize out of this carpet, lift over a rise and disappear, an occasional Merlin hunts from the four-foot pinnacle of a stunted fir. Everything knows how to be low, how to hug the rock and hunch against the wind. By the time you get to Mistaken Point, you will have already grown accustomed to looking down and looking closely, especially if it happens to be foggy, which is likely. You will also, probably, have had enough experience being buffeted by wind to appreciate the ecosystem's preference for a horizontal lifestyle.

The fossils are printed (although 'embossed' would be the more appropriate equivalent to the geologists' 'epirelief') on flat tilted beds of sedimentary rock right next to the sea. Some of them resemble ostrich feathers, some resemble elongated spindles, and some Bradgatia—are bushy. One of the Ediacarans, an unusually long frond called Charnia wardi, has been named after the Ward family from nearby Portugal Cove South. It was Catherine Ward and her son Brad who, having spotted two Americans trying to steal specimens from the fossil beds using a diamond saw, blockaded the road and called the RCMP. And it was Brad (who is now a geophysicist) who found the best example of the species that now carries the family name. Charnia wardi has since proven to be the oldest complex organism (that is, multiply-celled) in the fossil record, and the tallest of the Ediacarans.

Like all fossils, the Ediacarans are, in Christopher Dewdney's phrase, "pure memory," and seem to call, life form to life form, across 575 million years of evolution and geological transformation. It's as though the usually mute siltstone were sending semiotic signals. These animals (or 'animals') were soft-bodied stalks, connected to the ocean floor by hold-fasts, like kelp, living a life perhaps similar to today's jellyfish. It has been speculated by some scientists that they may have existed in a symbiotic relationship with primitive plants, perhaps as a plant-animal hybrid. They preyed on no other creature and, it seems, were not preyed upon themselves. Because of this, the period has been nicknamed The Garden of Ediacara by Mark McMenamin, apparently existing before predation, when symbiosis rather than predation was the order of the day. And, it also seems, they perished abruptly after 50 million years, when some creatures developed the canny notion, which has held sway ever since, that a quick way to nourish yourself is to eat somebody else. That, for the Garden of Ediacara, would have been the equivalent to the Fall. Enter the era of claws and shells.

Listening in on such geopoetry as, in the spirit of Harry Hess, I venture to call it, one feels one's thinking stretch as it takes on these remote possibilities. That stretch is, I think, not only epistemological (having to do with knowing) but ontological; it involves wonder at the manifold possibilities of being in general, and these beings in particular. Within a purely rational or analytical context such theories crave closure, desire to resolve into fact. The poetic frame permits the possible (I'm thinking of the sense in which Richard Kearney develops the concept) to be experienced as a power rather than a deficiency; it permits the imagination entry, finding wider resonances, leading us to contemplate further implications for ourselves. For although we are palpably here, our presence is no less a remote possibility in the long accident-riddled course of evolution than is that of the Charnia wardi and other Ediacarans embossed on the rock.

Today, at Mistaken Point, you can caress the rock with your finger and read their unreadable lines like Braille. You can trace the line between the fossil-bearing siltstone and the petrified volcanic ash that, ironically enough, both killed and preserved them, their assassin and archivist. These particular creatures were living off the coast of Gondwana when the volcano erupted, sending a cloud of ash high into the air, to be carried over the ocean. (Think of the extent of the fallout from Mount St. Helen's.) Eventually the ash particles settled into the water, smothering the Ediacarans under a soft grey cushion. Here and there in the ash layer (now a thin gritty black film) you can see bits of pink feldspar that crystallized out of the magma in the original eruption. You might have the sense, as I have had, that the fossils have been unveiled, as though some intentional hand were eroding the ash to reveal the beautiful fronds and disks beneath. Heidegger's term for beauty, *unconcealment* or *aletheia*, seems almost literally enacted by geologic forces. It even seems as though the slab on which they appear had been pulled from the other strata on the adjacent cliff like a drawer pulled open in a morgue for the corpse to be identified—an image that no doubt springs to mind due to my unhealthy predilection for cop shows, where it is a mandatory scene. If you're lucky, very lucky, it's sunny, and it's evening, so the slant light emphasizes the slight rise of the figures from the rock (their epirelief) and calls them to special eloquence, along with the deep nostalgia that dusk always lends its subjects. Pure memory. It is 570 million years ago on the other side of the Iapetus Ocean, an ocean that by the end of the Palaeozoic will have closed like a slow gigantic wink, along the continental shelf of Gondwana, the parent continent of both Africa and the Avalon Peninsula. Slim creatures sway at different heights in the tide, giving and taking from the water, existing in a world without predators. It is also, say, a Tuesday in September on the southern tip of the Avalon; the sun is setting; you'd better get going if you want to reach your car before dark.

In a geopoetic experience, like the imagined field trip to Mistaken Point, both elements, the 'geo' and the poetic, give something, and both, I think, inhibit or counteract a tendency in their partner. I am thinking here, as may be obvious, of a simpler version of the complex inter-relations between members of a symbiosis. Geology, or broadly speaking natural history of any kind, brings the rigour of the scientific frame; poetry brings the capacity for astonishment and the power of possibility, or, perhaps more accurately, legitimizes them. Geology inhibits the tendency, most common in romantic poets, to translate the immediate perception into an emotional condition, which is then admired or fetishized in preference to the original phenomenon—fossil, bird, lichen or landform. For its part, poetry cultivates the astonishment that naturally occurs in the presence of such marvels. As Adam Zagajewski says, poetry allows us "to experience astonishment and to stop in that astonishment for a long moment or two." By doing so it counteracts the tendency, perhaps most common in scientists in the grip of triumphalist technology, to reduce objects of contemplation to quanta of knowledge. Astonishment, humbling our pride in technique, impedes its progress into exploitation and appropriation. In the astonished condition, the other remains other, wilderness remains wild. Robert Hass, in *Time and Materials*, makes a cogent observation, which speaks to Zagajewski's idea of poetry. Interestingly, for the would-be geopoets among us, Hass writes these lines in response to an old lava field:

> It must be a gift of evolution that humans
> Can't sustain wonder. We'd never have gotten up
> From our knees if we could.

Hass makes a good point here, fine, contemplative nature poet that he is. If we could sustain wonder, we'd probably all have been devoured by sabre-toothed tigers long before *homo erectus* could evolve into *homo sapiens*; we'd be gawking at the marvel of the hairy mammoths and neglect entirely to slip our clever clovis-pointed spears between their massive ribs.

Nevertheless, speaking as a human existing in the outflow of the scientific revolution, living in a period of technological mania, I can't help but feel that we would have benefitted from spending more time on our knees, rapt, attending to the being of the other rather than classifying, analyzing, controlling, exploiting, and generally rendering the world as standing reserve available for our use. This is one of the ways in which poetry—any poetry—is always political and subversive: it uses our foremost technological tool, the ur-tool that is language, against itself, against its tendency to be the supreme analytic and organizing instrument. In poetry, language is always a singer as well as a thinker; a lover as well as an engineer. It discovers and delights in its own physical being, as though it were an otter or a raven rather than simply the vice president in charge of making sense.

Well, perhaps my characterization of the geological and poetic elements as symbionts is more of a hope than an observation, a self-serving attempt on my own part to integrate diverse bits of my cluttered life. But it does seem worthwhile to entertain the possibility that the two elements may, at least in isolated instances, feed, and feed upon, one another, and not just inhibit their respective excesses. When the intense experience of poetry, that momentary lyric peak, diminishes, we can turn to a more empirical attitude with a trace or memory of it persisting in our approach. The afterlife of wonder might well persist as a spirit animating the frame of knowledge. And likewise, the thirst to know, which has since Aristotle been recognized as fundamental to human sensibility, might be understood as an accelerant to poetic attention, rather than—as is usual—an aesthetic turn-off. The impact of the Ediacaran fossils is not diminished by a recognition of their place in the evolution of early life of the planet. In fact, I venture to think that such scientific reflections may serve to extend the condition of wonder from its peak epiphany into everyday existence. We might find it spreading from exceptional instances, like a trip to Mistaken Point, to the nondescript rock in my back yard,

Photo: Phoebe Cohen

which turns out to have travelled here from its birthplace in a volcano on the continent that became today's Africa.

This brings us, with more of a lurch than a glide, to the second time period named in the title, the Anthropocene Epoch. The Anthropocene has been proposed, though not yet officially recognized by the International Union of Geological Sciences, as the name for the epoch in which we are now living, an epoch characterized by the profound effect on the earth's systems of one species: *anthropos*, us. If generally accepted it would succeed the Holocene, the epoch that has extended from the ice ages (the Pleistocene) to the present. The date proposed for its onset differs from thinker to thinker, some placing it at the industrial revolution, others spotting the writing on the wall as early as the discoveries of agriculture or fire. Whatever the starting point, it is judged that the innovative technologies of *anthropos*, leveling forests, making cities, producing networks of roads, eliminating some species and domesticating others, have altered the workings of the planet's cycles in a way analogous to an ice age or a collision with an asteroid. Most tellingly, we've been digging up fossilized organisms and burning them, effectively turning earthbound carbon into atmospheric carbon, drastically altering the climate, as has occurred at other times in the earth's history when a greenhouse effect has come about from other causes. As Dean Young (not someone you might think of as an environmental poet) puts it, "Somehow / we've managed to ruin the sky / just by going about our business, / I in my super XL, you in your Discoverer." Writing, prophetically, in 1973, Christopher Dewdney observed that the effect of all the highways and associated fossil fuel emissions would be a kind of renaissance for the old Mesozoic atmosphere in which the plants originally grew.

The philosopher Emmanuel Levinas has given us a definition of European culture that resonates, in a sinister way, with the naming of the new epoch. "Culture," he says, "can be interpreted as an intention to remove the *otherness* of Nature, which, alien and previous, surprises and strikes the immediate identity which is the same of the human self." As an intention that converts the otherness of nature into the sameness of humanity, Levinas' culture sounds alarmingly like Calgary, eating its way steadily toward the Rockies, converting foothills into dismal suburbs of itself. It is against such reduction to the Same that poetry works, introducing otherness, or wilderness, into consciousness without insisting that it be turned wholly into knowledge, into what we know, what we own. Within poetic attention, we might say, what we behold is always "alien and previous," whether it's an exceptional fossil or an "ordinary" rock or chickadee. In poetry there is no "been there, done that;" everything is wilderness. The arrival of the Anthropocene would be an acknowledgement that the intention of culture, as Levinas sees it, has been all too richly realized, that there is little hope for an other that remains other, for wilderness that remains wild. It implicitly acknowledges that there will be no epoch called the Gaiacene, even though the concept was developed and maintained during the last century. In fact, the author of the concept, James Lovelock, is among the least optimistic of the earth scientists contemplating climate change.

Now, there actually is a way that culture has addressed nature during the last two centuries that is not exploitive or consumptive, at least on the surface, and that is Romanticism. Surely this must be reckoned a good thing, since it does not lead, like technology, to a reduction of the natural to either raw material or product. This is true. But Romanticism (of course, I am indulging in a lavish generalization) preserves the other not by respecting its otherness, but by welcoming it into the Same as a form of humanism. Nature as the kindly, pedagogical nurse in Wordsworth's poetry leads us to hear, not some "alien and previous" harmonies but the "still, sad music of humanity." No less than the technological mindset, Romanticism converts the other into the Same of the human self, but by a soft and seductive path, the generous extension of citizenship rather than violent reduction to utility. One thinks of certain Americans who praise our national character by announcing, generously, that it is the same as theirs.

I was struck afresh, recently, by the famous stolen boat passage from Wordsworth's *The Prelude*. You probably recall it, but let me summon it to mind in some detail, since it raises, quite insightfully, I think, the issue of wilderness, or the unassimilable otherness of the other. As he remembers it, Wordsworth "borrowed" a boat one evening and rowed out on Lake Windermere, getting far enough from shore that the perspective altered and a distant peak, occluded when closer in, suddenly loomed:

> . . . The huge cliff
> Rose up between me and the stars, and still
> With measur'd motion, like a living thing
> Strode after me.

Panicked, he beat a hasty retreat back to shore and, in the days following, was deeply troubled, for it was not the usual contact with Nature, which, as mentor and pedagogue, guided his development in humanistic ways. He had suddenly experienced wilderness-as-other, remorseless and terrifying. Of course, there is always a touch of terror in the experience of the sublime, but the young Wordsworth had received an overdose, and it would leave him at a loss to bring it into harmony with his earlier understanding of Nature:

> . . . and after I had seen
> That spectacle, for many days, my brain
> Worked with a dim and undetermin'd sense
> Of unknown modes of being; in my thoughts
> There was a darkness, call it solitude
> Or blank desertion, no familiar shapes
> Of hourly objects, images of trees
> Of sea or sky, no colours of green fields;
> But huge and mighty forms that do not live
> Like living men mov'd slowly through my mind
> By day and were the trouble of my dreams.

In this wonderful passage, the scrim of humanism is torn aside, to be replaced by unknown modes of being, huge and mighty forms that do not live like living men. The experience has revealed the wilderness lurking, like the distant peak, behind and within the idea of Nature, not only resisting the domesticating power of mind (refusing to become the same of culture) but indeed going on the offensive—pursuing, watching, troubling his dreams. Recall Margaret Atwood's query of the lady beholding the apparently innocent relief map of Canada: "Do you see nothing/ watching you from under the water?"

It is interesting to speculate, from a Canadian point of view, on what might have occurred to the young Wordsworth had he been unable to escape that wilderness experience, but been forced to live within it. What if he'd had to live with the "alien and previous" other rather than a Nature that endorsed human values? Well, I think Earle Birney provides the answer in his classic poem "Bushed," a poem that identifies the psychosis we have come to recognize as the consequence of an overdose of wilderness overwhelming the European consciousness. It is notable not only as a cautionary tale for all crypto-Wordsworthians, and not only because of its toothed imagery and curt music, but also because it articulates a crucial ambiguity at the heart of this breakdown. Is Birney's Romantic protagonist, having had his illusions about the sublime wrecked by the alien wilderness, about to become a nutcase or to enter the state of privileged consciousness that we would call, were we Native Americans rather than displaced Europeans, shamanistic? Birney doesn't say:

> then he knew though the mountain slept the winds
> were shaping its peak to an arrowhead
> poised
> And now he could only
> bar himself in and wait
> for the great flint to come singing into his heart.

For geopoetry to work, then, it must avoid Romantic humanism, despite its consider-able uplift and charm, and acknowledge the alien and previous character of the wilderness up front. The astonishment of poetry is right next door to being petrified—as the young Wordsworth and Birney's protagonist discovered. Wilderness does not endorse us as humans; it includes us as mammals.

Entering the Anthropocene, it seems, places our gifted but difficult species in a spotlight. At one time, i.e. the Enlightenment, such a focus seemed the illumination that relieved an oppressive darkness, enabling humankind to know itself and exercise fully its intellectual capacities. Now that spotlight may be more analogous to the headlight in which the deer is caught. But despite the dire events and portents that led to it, news of which assaults us daily, I believe there is a positive side to the nomination of the Anthropocene. All poets take naming seriously, aware that such baptisms into language carry enormous potential power. Language is, as John Steffler observed in his recent E.J. Pratt lecture, the first technology that we, the technological animals, have developed; it is the technology upon which all others depend. Naming might be said to be its first move in the conversation of the "alien and previous" into the familiar, accessible and manipulable Same. But the naming of the Anthropocene differs from others in at least two ways. One is that it is partly a negative recognition. Usually, if you get something named after yourself, or you name an organism, like *Charnia wardi*, after its discoverers and stewards, it's an honour. In the case of the Anthropocene, naming is an acknowledgement of responsibility and, in some measure, guilt. Although I do not rule out the possibility that there are folk out there cheering the advent of the epoch as a victory for our side, maybe even in some perverse way the material triumph of humanism, most will regard it as an act akin to naming atrocity atrocity or genocide genocide. Negative recognition, as with the familiar practice at A.A. meetings of identifying yourself as an alcoholic when you rise to speak, can become empowering. Among the recent works of Canadian poets in this vein, I would cite Dennis Lee's remarkable books *un* and *yesno*, with their torqued contorted techno-speak, as well as Pierre Nepveu's *Mirabel*, which brings poetry to witness the replacement of a pastoral landscape with a useless airport.

The second thing I have to say about the naming of the Anthropocene, and the last item in this *breccia* of an essay, regards its function as an entry point into deep time. If we think of ourselves as living in the Anthropocene Epoch, we realign our notion of temporal dwelling. Generally, time is viewed in relation to humanity's place in it, and consists of a present, where we live, and a recent past called history, which is felt to be important for informing the pres-ent and helping us understand ourselves better. When we speak of the past with reverence or chagrin, it is this shallow past we mean. Before history there is a vague distant past called prehistory, comprised of a jumble of relics and catastrophes, dinosaur bones mixed with clovis points, missing links, Lucy and The Flintstones cohabiting in the caves of Lascaux, Australo-pithecus confused with archaeopteryx, and the whole *mélange* construed as a sort of amniotic stew from which we, the Master Species, miraculously emerged. The name "Anthropocene," paradoxically enough, puts a crimp in this anthropocentrism, making the present a temporal unit among other epochs, periods and eras. If we think backward from the Anthropocene we encounter, like rungs on a ladder, the Holocene, the Pleistocene, Pliocene and Miocene epochs, and by this point there are no humans around, or even representatives of the *homo* genus, and we realize that the ladder extends back through periods and eras to the Ediacaran, and that even at this point we've covered only half a billion of the planet's four and a half billion years. On the one hand, we lose our special status as Master Species; on the other, we become members of deep time, along with trilobites and Ediacaran organisms. We gain the gift of de-familiarization, becoming other to ourselves, one expression of the ever-evolving planet. Inhabiting deep time imaginatively, we give up mastery and gain mutuality. ∎

Selective Sources

Atwood, Margaret. 1968. "At the Tourist Centre in Boston." In *The Animals in That Country*. Oxford: Oxford University Press.

Birney, Earle. [1951] 1975. "Bushed." In *The Collected Poems*, Vol. 1 Toronto: McClelland & Stewart.

Bjonerud, Marcia. 2006. *Reading the Rocks: The Autobiography of the Earth*. New York: Basic Books.

Dewdney, Christopher. 1983. *Predators of the Adoration: Selected Poems, 1972–1982*. Toronto: McClelland & Stewart.

Hass, Robert. 2007. *Time and Materials: Poems 1997–2005*. New York: Ecco, 2007.

Levinas, Emmanuel. 2000. *Entre Nous: Thinking-of-the-Other*, trans. Michael B. Smith and Barbara Harshaw. New York: Columbia University Press.

Stanley, Steven M. 2004. *Earth System History*. New York: W.H. Freeman.

Wordsworth, William. 1805. *The Prelude*.

Young, Dean. 2002. *Skid*. Pittsburgh: University of Pittsburgh Press.

Zagajewski, Adam. 2002. *Another Beauty*. Athens: University of Georgia Press.

Thanks to the staff at the Interpretation Centre in Portugal Cove South, especially Richard Thomas and Julie Cappleman, to Paul Dean at the Johnson Geocentre, and to provincial palaeontologist Doug Boyce. All have been very helpful and informative. Such foolish errors and extrapolations as appear are my own doing.

SLAG: A BY-PRODUCT OF EXTRACTED ORE

"IN SUDBURY, SLAG, ONCE HARDENED INTO ITS SOUVENIR-WORTHY FORM IS DISTRIBUTED AS THE MATERIAL FOR DRIVEWAYS THROUGHOUT THE CITY."
— ETIENNE TURPIN, *FROM "LIKE EXISTENCE: EXHUMING THE DRIVEWAY"*

IMAGE: LISA HIRMER, *UNTITLED*, FROM *SUDBURY SLAG*, 2012

Bill Gilbert

# 5. MODELING COLLABORATIVE PRACTICES IN THE ANTHROPOCENE

Claire Cote, *Untitled*

In 1999, the University of New Mexico embarked on an extended experiment in art practice. Over the ensuing twelve years we have developed the Land Arts of the American West program as a model for a place-based education in the arts to prepare our students for the rapidly changing environmental and social context they will enter upon graduation.

Our program has developed in a period marked by a dawning awareness that the collective activities of our species have impacted the planet over a sufficient time period to qualify as a geologic epoch, the Anthropocene. With this realization comes a daunting responsibility. The arts must now participate in our collective response and contribute to a change in our narrative as a nation and species if the Anthropocene is to extend for thousands of years into the future.

To be successful we are going to need a perspective that encompasses the expanse of planetary time, not the fleeting moments of pop culture. In our effort to confront the problems we face, the arts can model a new cooperative/collaborative approach that will supplant the current individualistic paradigm. Our attempts to address the implications of an Anthropocene Epoch will require a shared multicultural and interdisciplinary perspective that is based in direct engagement with the physical planet.

Land Arts of the American West, in its current role as the field program in our Art and Ecology curriculum, is designed to provide students with an experience based in direct contact with the environmental and cultural aspects of our region and a collaborative response to existing conditions/problems.

UNM is situated in the social context of the city of Albuquerque and the environmental context of the high deserts of the American west. As such, we are located at the cultural nexus of the major migrations of human populations that have driven habitation on this continent:

the north-south migrations of first Native and then Hispanic cultures and the east-west migrations of northern European cultures. In ecological terms, we are placed in the exposed topography of the deserts of the Americas where the axis of the geologic and the geographic meet. The horizontal axis of geography, the surface of the planet, intersects the vertical axis of the geologic, strata upon strata of mantle surrounding a magma core. This intersection is eminently visible in natural events throughout the west including riverbanks, landslides, and escarpments and also in the layers revealed by man's activities in constructing rail lines, highways, and mines.

Each fall Land Arts at UNM sets off from our home base in Albuquerque to navigate the geography of the west. Skimming across the surface of the planet, moving freely through space as a nomadic band of art hunters and gatherers. We cover vast swatches of the desert landscape. Students experience first hand the range of interventions in the planet's surface made by Native American and Hispanic cultures, contemporary Land Artists, and the military industrial complex developing a sense of how humans have altered the environment across a 6,000-8,000 year period.

Moon House, Cedar Mesa, Utah. Photo Credit: Bill Gilbert

As the successive years of the program have been laid down one on top of the next, we have become aware of a slow accretion of experience. We have now been to Spiral Jetty eight times in an annual visitation each fall. At our first encounter, the Jetty was almost entirely submerged with just a few rocks poking through the surface of the lake providing a faint outline of the spiral. We have seen it slowly emerge covered in a bright white salt crust, turn dingy grey and land-locked only to subside once again as the waters of Salt Lake have continued their cycle of rise and fall. Our years of contact becomes a lens through which to grasp the Jetty's forty-year cycle of creation, submersion, and emergence that, in turn, points us towards the deep geologic time of ancient Lake Bonneville slowly receding to leave the current Salt Lake, undoubtedly destined to rise once again in a distant future.

Our focus has now changed in response to these investigations in place. We have moved from the grand survey, an art tourism approach, to one of engagement and commitment, building relationships with a limited number of sites and communities. No longer skimming the surface of the west, we have accepted the implications of the Anthropocene and have begun to build sustainable roots by drilling down into specific environmental and social places.

Our sense of what our work is and how it operates in the world has changed, as well. Originally, the focus was on the aesthetics of marks left on the surface of the planet by Native, Hispanic, and Anglo cultures; pictographs, petroglyphs, earthworks, dams, mines, etc. Faced with the environmental and social upheavals brought on by global warming we are now investigating a different focus for art practice; one based in a collaborative dialogue with place and community.

In some cases this means using our skills as artists to change the narrative around existing cultural interventions. In 2009, the Land Arts program worked at the Center for Land Use Interpretation's base in Wendover, Utah to create a work entitled *A Hole To China* based on Lucy Raven's film animation entitled *China Town*. After a group trek with CLUI director Matt Coolidge and guest artist Lucy Raven to the Bingham Copper Mine outside Salt Lake to see one of the largest man made holes on the planet,

Bingham Mine, Salt Lake City, Utah. Photo credit: Jeanette Hart-Mann

we decided on a project that would articulate the particularities of the connection forged by copper mining/smelting between Ruth, Nevada and Tongling, China. Our intervention, installed by the students, consisted of interpretive signs and pamphlets.

A Hole to China, Land Arts collaboration with Lucy Raven
Photo Credit: Jeanette Hart-Mann

Adapting the aesthetics and forms of US Government signage, we offered an alternative interpretation of the mining activities based in the saying from childhood that we are "digging a hole to China."

In other contexts this has meant working with community members in projects to build environmental and social sustainability. In recent years, Land Arts has returned to CLUI's South Base station to collaborate with Matt Lynch and Steve Badgett of Simparch at *clean livin,*' an experiment in sustainable living on the Wendover Air Force base. In this case,

*clean livin,*' Simparch, CLUI southbase, Wendover, Utah. Photo Credit: Steve Badgett

the Land Arts program contributed to Simparch's experiment in sustainability by building a water catchment, planting bed, a human-activated composter, mobile shade structure, and a solar oven.

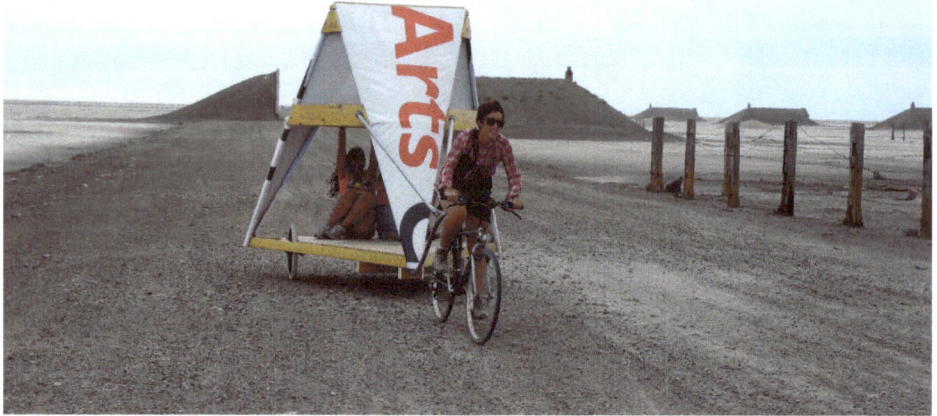

Mobile Shade Structure, Simparch/UNM Land Arts program collaboration CLUI southbase, Wendover. UTPhoto credit: Jeanette Hart-Mann

In the Buena Vista Neighborhood of El Paso, TX we have established a partnership with El Centro Artistico y Cultural (CAYC) through its director Roberto Salas. Land Arts is involved in a long-term commitment to assist the neighborhood efforts to improve the quality of life for its residents. In collaboration with the CAYC and the Buena Vista Neighborhood Association (BVNA), Land Arts first created a 200' by 6' mural in 2009 out of recycled marble counter tops to enhance the entrance to the *barrio*. In 2010, we helped community members develop a

Barrio Buena Vista Master Plan, Centro Artistico y Cultural/UNM Land Arts program collaboration, El Paso, Texas. Photoshop: Bethany Delahunt

master plan for Buena Vista that included an initiative to protect the neighborhood from Department of Transportation efforts to annex neighborhood land to expand I-10, and a proposal to add wetlands currently belonging to Cemex Corporation to the community park.

This year we responded to a long neglected community request by designing and constructing a bus stop for the neighborhood. Older community members now have a shaded place to sit as they wait for the bus in the 100-degree plus heat of El Paso summers.

*Bus Stop*, Centro Artistico y Cultural/UNM Land Arts program collaboration Barrio Buena Vista, El Paso, Texas. Photo credit: Nina Dubois

We are currently working with CAYC, BVNA, the City of El Paso, and UNM biologists on a grant to the EPA that would provide funds to restore the adjacent wetlands and train community members to monitor environmental conditions in the neighborhood.

Each of these projects by themselves will do little to alter the course of the environmental and social change. They do, however, model a fundamental shift in how art as a practice is defined at the University of New Mexico and how Art and Ecology students are being prepared to engage with our culture as we strive to establish a sustainable society and extend the run of the Anthropocene Epoch. ■

Erika Osborne

## 6. EXPOSING THE ANTHROPOCENE: ART AND EDUCATION IN THE "EXTRACTION STATE"

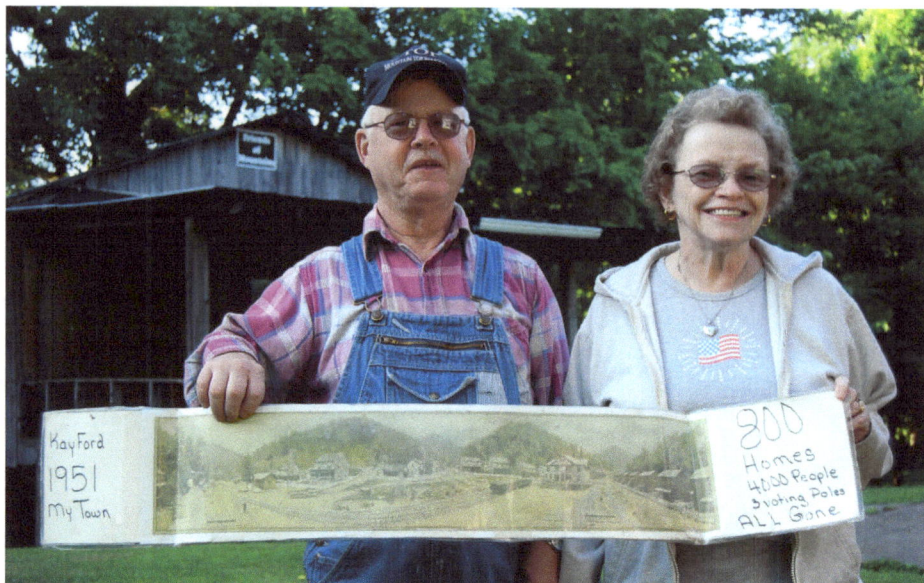

Larry and Carol Gibson with a photograph of a populated Kayford circa 1951.
Photograph: Erika Osborne

In the age of the Anthropocene, human impact on a geologic scale infiltrates nearly every aspect of contemporary life. The immense breadth and depth of the changes to the planet that have been caused by this impact have affected my work and my pedagogy as an artist and educator interested in connections between culture and environment.

Although I have always been interested in the Anthropocene, my fascination with its artifacts grew when I accepted a teaching position in the School of Art and Design at West Virginia University and moved to "The Mountain State"—or what West Virginia's former governor and now senator, Joe Manchin, has lovingly called "The Extraction State." Here, it is obvious that not only have the surface and ecology of mountains changed because of deforestation, but entire topographies have morphed in a geologic instant as a product of large-scale mountaintop removal (MTR) mining. This type of mining is a quick, relatively inexpensive process in which the summits of mountains are blasted away to expose coal seams, and the overburden is dumped into adjacent valleys. This anthropocentric practice has come to define the state and has created a battleground, pitting local communities and the mountain ecology against an ever-growing, global need for cheap energy. In mined areas of Appalachia, the biodiversity of some of the oldest mountains in the world is obliterated instantly, habitats are immediately lost and clean water sources are buried or left highly contaminated. In addition, the local people often suffer from serious health issues affecting their lungs, kidneys, hearts, and nervous systems. They also suffer economically as property values drop, businesses close, and economic diversity is lost.

In addition to defining the state in which I reside, this form of corporate colonialism and its geologic, ecologic, and biologic effects has come to play a large role in my pedagogy as a professor here. As part of two place-based field courses I have developed, *Art and Environment* and *Place: Appalachia*, students spend from one to three days on Kayford Mountain, an MTR site in the southern coalfields of West Virginia, with local activist Larry Gibson. Larry is evangelical in his speech, preaching the need for action to halt MTR mining. The effect of listening to him speak, coupled with seeing mountains literally moved by giant machines, is profound and visceral for the students. The three-hour van ride is much different on the return trip than the trip to Larry's place, the latter being full of music and chitchat, the former being almost eerily silent. When asked what they think of the experience, many students find themselves without words as they try to make sense of what they have witnessed and what it means for them. Although silent at first, the experience ultimately leads to discussions about how to process the encounter not only as an artist, but also as a member of a contemporary culture that survives on the energy created by such practices.

Most of the students are from West Virginia or the surrounding region, and grew up with the coal industry. However, until their trip to Kayford Mountain, many of them have no idea of the extent to which the industrial practice impacts the environment and the people. Students participating in *Place: Appalachia* are invited to create art about this issue while on-site during their three-day stay on Larry Gibson's property. They work with what supplies they bring with them and what they can find on site. They hear daily blasts from MTR mines in the area, explosions that sound like distant claps of thunder. Black bears, displaced by the deforestation that accompanies the mining, come through camp on a daily basis.

Painting of a processing impoundment in Sylvester, WV.
Painting: Carrie Grubb

The Art and Environment students spend only one afternoon on Kayford Mountain, but they too are able to experience the destructive violence of the MTR practice. Students from both courses are able to project what they see at Kayford onto the tree-covered mountains they see in the surrounding area, many of which are already slated to be mined. Students also interact with locals who are both for and against current coal mining practices, people who have hunting cabins on Kayford Mountain or who live in the adjacent Coal River valley. Although Larry's personality keeps students smiling throughout their visit, the initial artistic reactions are typically shock and anger and the art that is made is often heavy in mood and dark in color. On site, many students see the issue as black and white. It isn't until they return to Morgantown and start to seriously process what they have experienced that they begin to see the contradictions and nuances surrounding the practice. Grey infiltrates the black and white every time they turn on a light switch.

A portion of the Massey mining complex, Kayford Mountain. Photograph: Elizabeth Ruwet.

It is this daily confrontation that allows the experience students have on Kayford Mountain to change how they navigate the world. At a minimum, students move through their days more consciously, turning their thermostats down and their light switches off. However, for many, the visceral impact of what they see and hear on Kayford resonates deeper into their life practices. Many continue to create work about MTR mining itself, often diving deeper into its intricacies. The work that surfaces is exponentially more sophisticated than their initial attempts as they continue to research, to revisit the coalfields and to acknowledge the paradox

Kayford coal. Photograph: Caitlin Ratliff

in which they find themselves. In addition to creating larger bodies of work that address the mining practice, many students take it upon themselves to let friends and family members, many of whom are ignorant of the issue, know what is going on. They do this by distributing information via social media, encouraging friends and family to watch documentaries on the subject, taking people to Larry Gibson's to see the mining firsthand, and even by creating t-shirts with anti-MTR slogans to wear on campus. Still others extend their engagement with this destructive mining practice into social and environmental activism.

As artists, these students have donated artwork as a means to raise money for foundations that help fight the mining onslaught in local communities. They have also used their skills as artists to create posters, signage and t-shirts for rallies. In addition to donating their services, many students have become entrenched with the community of activists who are trying to raise awareness about the issue. Students have participated in many of the non-violent *Appalachia Rising* marches, including marches on Washington D.C. and, most recently, a 50-mile march to Blair Mountain, a historic sight marking the famous civil uprising of coal miners against coal operators, police, and eventually the United States Army. As Blair Mountain is being threatened by MTR mining, students have been willing to risk arrest to appeal to the government to work toward a sustainable future for Appalachia. One that preserves the environment, promotes a diverse economy, and creates healthier communities.

Students wearing their activism. Photograph: Erika Osborne

This expansion of art practice into life practice belies a deeper shift in the minds and hearts of the students. No longer do they think of themselves as merely being subject to geologic forces in contemporary life. They now realize they are part of those changes. ∎

*–this essay is dedicated to Larry Gibson.*

"IF THE SOLAR STORM OF 1921, WHICH HAS BEEN TERMED A ONE-IN-100-YEAR EVENT, WERE TO OCCUR TODAY, WELL OVER 300 EXTRA-HIGH-VOLTAGE TRANSFORMERS COULD BE DAMAGED OR DESTROYED, THEREBY INTERRUPTING POWER TO 130 MILLION PEOPLE FOR A PERIOD OF YEARS."

- JOSEPH MCCLELLAND, DIRECTOR OF THE OFFICE OF ELECTRIC RELIABILITY AT THE FEDERAL ENERGY REGULATORY COMMISSION, MAY 31, 2011, HOUSE ENERGY SUBCOMMITTEE HEARING

IMAGE: CHARLIE BATES SOLAR ASTRONOMY PROJECT

# SECTION 2: SHIFTS IN THE MATERIAL CONDITIONS OF CONTEMPORARY LIFE

Seth Denizen

## 7. WHAT IS THE EXPONENTIAL?[1]

The strange thing about these exponential curves, is that the farther along the curve one projects the present, the shorter the time interval between successive points, until time all but stops, in the midst of an immense acceleration. This is the point at which the curve challenges us to imagine a world defined not by a time, but by a speed: one which cannot be merely an *extension of our own, a difference in degree, but rather something which takes on a difference in kind: another sea, another wind, another world at right angles to our own.* ■

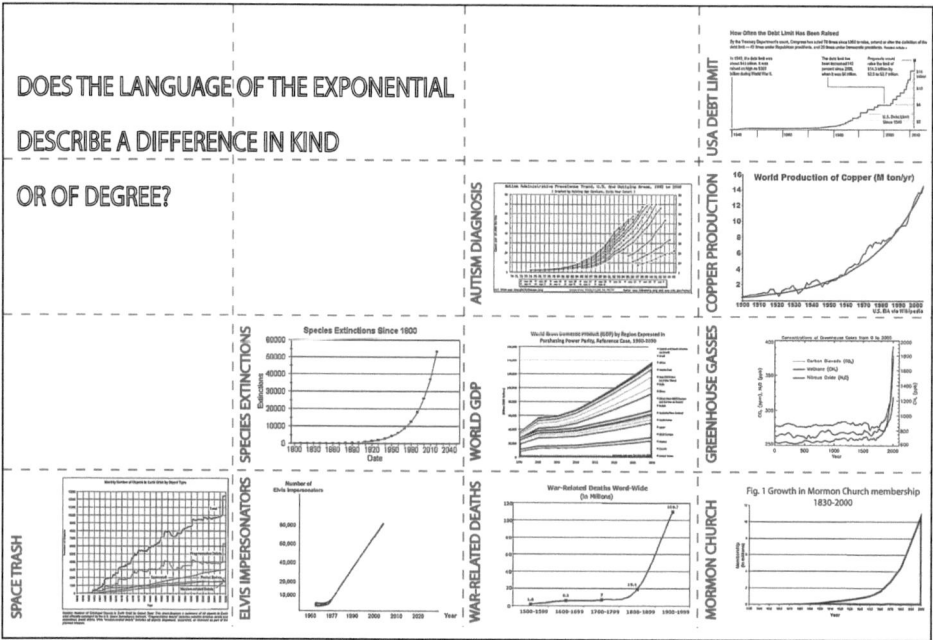

Does the language of the exponential express a difference in kind or of degree?; image Seth Denizen

---

1  From Seth Denizen, "Geopoetic Inconformities," paper presented at *The Geologic Turn: Architecture's Alliance* [symposium], Taubman College of Architecture + Urban Planning, University of Michigan, Ann Arbor, MI, February 10-11, 2012.

Ryan Thompson

## 8. THE DARK FLIGHT OF MICROMETEORITES

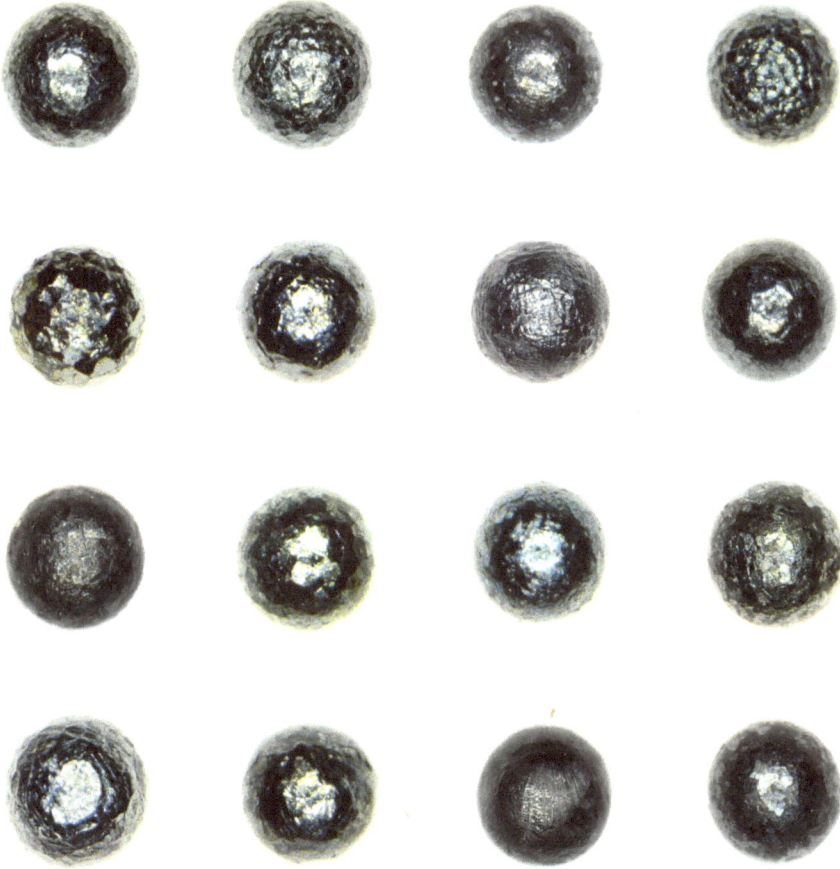

These 'turtlebacks', photographed by Jon Larsen of Project Stardust, are part of an ongoing study using typologies to differentiate micrometeorites from microscopic terrestrial spherules. Image courtesy: Jon Larsen, 2011.

Every day some 200 tons of extraterrestrial material enter the Earth's porous atmosphere. The largest of these objects, meteors, become giant fireballs with the ability to light up the daytime sky and can cause local, regional, or global destruction upon impact. Others become shooting stars, neither large enough to survive their fiery trip through the atmosphere, nor small enough to escape their fate. The smallest of these materials, however, make it to the surface of the Earth as micrometeorites without much in the way of fanfare. No fiery explosions in the sky. No damage or destruction. Just a silent fall to Earth.

By definition, these micrometeorites are generally less than 1mm in diameter—literally dust-sized. However, what they lack in theatrics or destructive power, they make up for in numbers. While a meteorite impact that has the ability to cause regional destruction may occur once every 10,000 to 100,000 years,[1] micrometeorite "impacts" occur millions of times *daily*. Estimates

---

1  See the "Torino Impact Hazard Scale," *Near Earth Object Program, NASA* available at: http://neo.jpl.nasa.gov/torino_scale.html.

vary widely with respect to just how much of this material the earth collects, with some studies claiming that the annual influx weight of these objects may be upwards of 14,000,000 tons. Far more likely, however, are estimates in the 10,000 to 20,000 ton range.[2] On the low end of these estimates, the daily influx works out to some 27 tons per day, accounting for the majority of extraterrestrial materials on Earth.[3]

According to written records, humans have known about objects falling from the heavens for a few millennia at least, and perhaps as far back as 1478 BCE.[4] It wasn't until the end of the 19th century, however, that we began to study the meteoritic dust that falls to Earth as well. Scientists have learned much in the past century about the meteorites that do make it to Earth—their age, their chemical make up, the frequency of their falls, the threat they pose to life on Earth—but it seems as if we are just beginning to understand the significance of what they might tell us about our place in the geologic now.

*An unclassified metallic spherule (possible micrometeorite) on the tip of my index finger. Image Courtesy: Ryan Thompson, 2012*

In the years and months leading up to the first manned mission to the Moon, NASA scientists worried whether or not the Lunar Lander would find steady footing upon touchdown. The debate centered on just how much space dust was covering the surface of the Moon. With no atmosphere and little erosion, some hypothesized that the Moon had been collecting the very same micrometeorite materials as the Earth—but in far greater quantities. Enough, in fact, to create a layer of space dust 20 feet thick across the entire surface of the Moon. Concerns for the spacecraft and crew were eventually allayed, and upon touchdown the Lander settled in no more than a few inches of loose space dust.

As they fall to Earth, some evidence suggests that micrometeorites are slightly more

---

2 S.G. Love and D.E. Brownlee, "A Direct Measurement of the Terrestrial Mass Accretion Rate of Cosmic Dust," *Science, New Series*, 262.5133 (1993): 550–553.

3 S.E. Peal, "The Dark Tints of the Lunar Maria," *Journal of the British Astronomical Association* VII.7 (1897): 400–401.

4 Patrick Moore, *The Data Book of Astronomy* (London: Taylor & Francis, 2000).

attracted to the magnetic poles, but as a rule of thumb, they're landing everywhere. In the oceans. On the tops of mountains. In parking lots. In cups of coffee. They've been stepped upon and ingested. Buried and unearthed. Found and forgotten again. They're buried deep in the arctic ice, having landed thousands of years ago, and they're falling fresh right now.

For the past few decades the lay community believed that searching for micrometeorites was as simple as collecting rainwater or sweeping up road dust and examining it under a microscope. While true that this approach may yield micrometeorite specimens, the authentication of these small metallic spherules as such is much more complicated. As it turns out, there are certain post-industrial processes, the flicking of a lighter or the use of a grinding wheel, among others, that have the ability to create morphologically similar objects.[5] In fact, in any successful rainwater or road dust sample one will also find scores of false positives, or "micrometeor*wrongs*." If with every flick of a lighter, we create a handful of micrometeorwrongs, it's not difficult to imagine these newly formed spherules existing along-side their micrometeorite counterparts for hundreds of years, or perhaps thousands, into the future. Scientists have therefore turned to collecting micrometeorites from pre-industrial melted arctic ice, and from the tops of mountains. To the layperson, however, it may all be the same. Micrometeorite or micrometeorwrong, we can imagine these dust-sized particles land-ing on Earth after being polished round from their multi-billion year journeys through space, ready for our rain-collectors, magnets, and microscopes, and ready for us to re-imagine our relationship to the cosmos as messier and more complex than we ever imagined it.

Despite floating through space for millions of years, upon arrival on Earth, micrometeor-ites become as much a part of our everyday, as they are a part of the origins of our solar system more than 4.6bn years ago. To compare our day-to-day human time scales to the time scale of micrometeorites in this fashion feels both humorous and humbling. That we might understand during an average lifetime what constitutes the space of billions of years, let alone thousands, seems quite impossible. But it's worth trying because doing so adds new perspec-tive to our perceptions of time.

Copernicus disproved the geocentric model back in 1543, but that doesn't mean, nearly five centuries later, that we can't still fall into old habits. Standing on the surface of the Earth, it's easy to summon the Ptolemaic system, imagining outer space as simply above us, stretching out from where we stand in a unique and privileged position in the universe. In this scenario, the Earth seems to occupy a much larger portion of the universe than it does. Imagining Earth as a micrometeorite floating through space may help put things back in proper perspective.

In the summer of 2011, the final NASA shuttle returned from space, ending a shuttle program that spanned more than four decades. While this doesn't signal an end to the U.S. space exploration program, it brings with it waves of nostalgia. While these feelings may not be entirely misplaced, it's worth remembering how porous our atmosphere is and how finite the space between Earth and the Moon is. The space beyond our upper atmosphere—"outer space"—is not just out there, but rather, is intertwined in our everyday. Micrometeorites remind us of this by shifting our perceptions of time, scale, and space. What we do not know about time and space is far greater than what we do, and we would do well to live with a greater respect for these effectively invisible events of the geologic such as the flight of a micrometeorite. ∎

---

5  Attilio Anselmo, "Observation of False Spherical Micrometeorites," *arXiv Archives, Cornell University Library* 2007: http://arxiv.org/ftp/arxiv/papers/0708/0708.4276.pdf.

Dredge Research Collaborative (Stephen Becker, Rob Holmes, Tim Maly, and Brett Milligan)

# 9. PACKAGING SLUDGE AND SILT

Geo-tubes being filled with sediments at a shoreline.

A massive, distended tube thirty feet in circumference and one hundred feet long lies on the bank of a drifting river, among the sludge and muck. A thick hose connected to a dredging barge is filling this black bag with the silt that it is sucking out of a nearby shipping channel. Once full, the tube will be tied shut and left in place. Fluids will slowly seep out, leaving dewatered sediments contained within the casing. This is the geo-tube.[1]

A geo-tube is like an oversized[2] sausage casing made of "geotextile," a synthetic fabric woven primarily from hydrocarbon-derived polymers.[3] When deployed, they are inflated by liquids, slurries, or sediments, depending on their intended use. Geo-tubes find their application where water meets land and where landscape meets industry. They are deployed along riverbanks, coastlines, in shallows, or wetlands. They have spread quickly thanks to their flexibility, speed of application, and cheapness.

---

1 Here we should distinguish between the generic term "geo-tube," an abbreviation referring to prefabricated geotextile tubes, a class of applied geosynthetics, and "Geotube® containers," a particular set of products manufactured and trademarked by TenCate Geosynthetics, a corporation owned by the Dutch materials technology company Royal Ten Cate. TenCate is the primary global supplier of geotextile tubes.

2 We use "oversized" advisedly: TenCate currently engineers Geotube® containers in circumferences of up to one hundred feet, with lengths close to five times that.

3 ASTM D4439, the international materials standard that covers geosynthetics, defines a geosynthetic as "a planar product manufactured from polymeric material, used with soil, rock, earth, and other geotechnical engineering-related material as an integral part of a human-made project, structure, or system."

Being little more than high-tech fabric bags, geo-tubes have no rigid or complicated parts. They fold away like industrial bedding when not in use; they take on the form of the raw material pumped inside them; they can be deployed in nearly any situation; and they are mass-produced. If the 20th century found much use and appreciation for the small and humble sandbag, the beginning of the 21st century ushered in the increasingly prolific application of its scaled-up successor. The geo-tube embodies shifts from hard strategies of concrete and steel to soft synthetics and engineered biologies, from public state actors to multi-national corporations.

TenCate Geosynthetics is the primary global supplier of geo-tubes. The business is a division of TenCate, an international materials technology conglomerate with divisions specializing in protective fabrics, space and aerospace composites, "advanced armour," synthetic turf, geosynthetics and industrial fabrics.[4] Their attractive marketing materials divide their geo-tube offerings into two categories. Geotube® Dewatering containers act as a repository for the oft-contaminated silts that are inevitable side effects of water treatment projects, the dredging of navigable channels, and industrial operations ranging from aquaculture to pulp mills. Geotube® Marine Structures act as rapidly-deployed structures, their flexibility and size making them ideal as temporary breakwaters, or for shoreline and island protection, to say nothing of rapid damming, underwater structure creation, sand dune maintenance, wetland creation and protection, jetties, groynes, dikes, and the production of whole new tracts of land. In the domain of coastal construction, there are few situations where geo-tubes are not applicable.

However, these marketing materials are relatively silent on the larger processes and forces that compel us to lay lines of enormous black plastic slugs across fluvial and littoral landscapes which have been eroding and accreting without human assistance for eons. They do not explain that the geo-tube is an ascendant tool in the larger systemics of anthropogenic erosion, what we refer to as the Dredge Cycle.

Diagram of the Dredge Cycle. The natural paths of erosion and sedimentary geology (white) are overlaid by the alternately accelerating and decelerating paths (orange) produced by various anthropogenic inputs. The diagram is divided between gravity-fed erosive paths on the right and forced uplift on the left. Distance from the center of the diagram abstractly represents time -- the farther the orbits are from the center, the slower they are. At the bottom, dredging is the key moment at which sediments shift from being under the influence of gravity to defying gravity.

4 Information on TenCate and TenCate's products is based upon TenCate's website, http://www.tencate.com, TenCate's marketing materials, and conversations with representatives of TenCate.

## THE DREDGE CYCLE

The Dredge Cycle is not limited to the activities of dredging machines (dredge as a linear act of industrial engineering), but encompasses a much wider range of anthropogenic sediment handling activities and practices. These act upon, feedback to, and destabilize one another, producing a quickened anthropogenic counterpart to conventionally defined geologic cycles.[5] This cycle can be split into two primary phases: anthropogenically accelerated erosion and the manufactured uplift of soil, silt, sand, and sediment. Like the rock, water, and wind cycles, it is ubiquitous, operating wherever humans intentionally or unintentionally move sediment.

The Dredge Cycle is geological in its processes and the material it cycles, but not in its duration. Its unintentional aspects are driven by the assumption of human control over erosive processes (hardtop surfaces prevent the absorption of water, causing accelerated erosion where that water flows; coastal structures alter current patterns, depositing more soil in one area while depriving another). Its intentional aspects are symptoms of economic and political systems that assume stasis in a landscape that was never static (homeowners fight to maintain beaches lest their front yard be washed out to sea; logistics firms demand uninterrupted access to shipping channels). The Dredge Cycle is a crass geology, run on a human timescale.

The Dredge Cycle consequently operates at an ever-increasing pace. The methods used to halt the forces that we set into motion are crude, and routinely throw sedimentary systems out of equilibrium. For instance, when a shipping channel is dredged to deepen it, this destabilizes the riverbed and the channel begins to fill at a faster rate than before the operation, which necessitates further dredging, which produces additional destabilization. Such cycles spiral inward, growing tighter and faster, across the breadth of the Dredge Cycle.

The primary forces driving the Dredge Cycle are urbanization, agriculture, and forest-clearing (which produce erosion on land and siltation in waterbodies), damming (which accelerates siltation in waterbodies), the need to maintain navigation channels for shipping (which requires removing sediment from waterbodies), and the desire to hold in stasis the shape of shorefronts (which typically requires the deposition of sediment to counter erosive forces, both natural and anthropogenic).

There are innumerable infrastructures, devices, machines, and products implicated in the Dredge Cycle: mechanical and hydraulic dredges, pipelines, barges, conveyor belts, trucks, bulldozers, backhoes, earthmovers, geotextiles, silt fences, turbidity curtains, containment booms, cellular confinement systems, sand bags, Hesco baskets, groynes, dams, and more. With this distributed mechanical apparatus, we humans are continually reorganizing the sedimentary geology of six of the seven continents.

Geo-tubes are relative newcomers to these systemics,[6] but they represent larger trends

---

5 By "conventional" or "classical" geology, we refer to models of geologic formation that exclude human influence on geologic processes and forms. Conventional geology can be contrasted with the movement among contemporary geologists to describe the "Anthropocene," a geologic era characterized by the discernable effect of human activities on the geologic record. The Anthropocene has been described in numerous recent articles in the general press, including a 26 May 2011 editorial briefing in *The Economist* entitled "The Anthropocene: A Man-made World," Elizabeth Kolbert's March 2011 "Age of Man" in *National Geographic Magazine*, and a 27 February 2011 editorial in *The New York Times* entitled "The Anthropocene."

6 TenCate's marketing materials indicate that the first installation of geo-tubes was in Europe in 1962. (This installation predates the trademarking of the name "Geotube®"—a trademark was filed with the US Patent and Trademark Office in 1995—but the technology appears essentially similar.) Though, like most industrial inventions, the geo-tube has a long international patent history, the most current patent held by TenCate Nicolon is US Patent No. 6,186,701, "Elongate Flexible Container," filed in 1997 and approved in 2001. The widespread use of geo-tubes in the United States similarly began in the 1990's and expanded throughout the first decade of the twenty-first century.

within the Dredge Cycle at least as well as any other material technology. The geo-tube is both a response to the increased speed and quantity with which humans encourage sediments to cycle across the surface of the earth, serving to slow coastal erosion, for instance, and a contributor to that process. The geo-tube makes it more convenient—faster, easier, and cheaper[7]—than ever to confine, place, and move sediment. This convenience breeds ubiquity; when something is fast, easy and cheap, the tendency is to do more of it. Geo-tubes are consequently one of the many forces driving the expansionary logic of the Dredge Cycle, feeding its tendency to bring ever-increasing quantities of sediment under the influence of human sediment-handling practices.[8]

Diagram of the expanding variety of geo-tube applications. Each use implies a particular architectural character for the installation and varying relationships between water, land, and geo-tube.

## A NEW GEOLOGIC VERNACULAR

Because of its flexibility and adaptability, the geo-tube is nearly omnipresent in the Dredge Cycle. Like a scaled-up merger between sandbags and silt fences, the geo-tube can be deployed as an erosion control measure that slows accelerated erosion through placement as a physical barrier. When used in **erosion control**, the geo-tube is itself filled with dredge

---

7  Krystian W. Pilarczyk, *Geosynthetics and Geosystems in Hydraulic and Coastal Engineering* (Brookfield: A.A. Balkema Publishers, 2000), 3.

8  It is not at all coincidental, given the role of the corporation in the growth of the Dredge Cycle, that this expansionary logic mirrors the growth logic of the corporation, in which innovation, standardization, and distribution serve to lower production costs, cyclically funding further innovation, increased standardization, and expanded distribution. The Dredge Cycle is what happens when corporate logic is applied to geology.

material harvested from clogged streams, rivers, and bays.  Alternatively, when employed as soft dams, where they, like all dams, act to trap sediment and induce **siltation**. The geo-tubes' porous synthetic skin allows for the **recovery** of sediment by permitting the passage of water while trapping suspended particles. This captured material can then be **refined**, or sorted into clean and contaminated components, which in turn can then be **reapplied** for a range of 'beneficial uses,' including the production of new land, submarine terrain, and landforms. This kind of cyclical feedback, a single technology contributing to the anthropogenic accelera-tion of erosion while simultaneously trying to decelerate or reverse those effects, is one of the most basic characteristics of the Dredge Cycle.

The geo-tube literally encapsulates the sublime materiality of the Dredge Cycle, as sedi-ment and water in slurried suspension are stuffed into geotextile casings. The Dredge Cycle is fundamentally composed of *wet stuff*: basic materials; ordinary sand, silt, clay, and water. While it can and should be understood as a highly abstracted set of networks and feedback loops operating on a global spatial scale, it should also be understood as a material opera-tion. It is the cubic yards of excavated soil downwashing across your backyard from the new construction three houses down in a rainstorm as much as it is globally networked processes like the expansion of the Panama Canal to accommodate the importation of goods from East Asia driving port expansions and dredging operations along the East Coast of North Amer-ica. Similarly, geo-tubes are always dirty: placed in muck, filled with muck, and, like muck, slumping and slouching into soft shapes, rather than following the precise angles of architec-tural geometry.

Geo-tube Dam's In Honduras. TenCate's Geotubes® have been deployed for "the construction of two dams in a river bed near a storage lagoon for mine tailings slurry in Honduras. The technology ensures storage, dewatering and filtering of polluted ground."[9]

9  See "TenCate Geotube® technology in Honduras," Tencate.com, Spring 2010: http://www. tencate.com/ 11839/TenCate/Corporate/en/Home/en-Home-Txtures/spring-2010/TenCate-Geotube-technology-in-Honduras.

The geometry of the geo-tube is no more natural than the clean modernist lines of the Hoover Dam. It is something else entirely, both post-natural and post-architectural. This seems entirely appropriate for an era in which we are freezing sediment-spraying rivers in specific configurations, like the Mississippi at Old River Control, or impounding the eroded sediments of entire continents behind vast concrete structures, like Three Gorges Dam. Our largest monuments are not pyramids and skyscrapers, but geologic impacts.

The Dredge Cycle is thus landscape design on a deliriously monumental scale, but unrecognized as an architecture. It instead remains the domain of logistics, industry, and engineering. Geo-tubes, like other geosynthetics and indeed nearly all of the infrastructure of the Dredge Cycle, are not usually placed or constructed with a particularly conscious aesthetic aim. Rather, they are part of the infrastructural vernacular of civil engineering's ordinary monuments. When they are used with conscious aesthetic intent, they tend to be hilariously unsuited to the cosmetic tasks they are asked to perform: resort-beach-bound geo-tubes literally whitewashed, for instance, to look sort of like pale sand, which only serves to make them more obviously awkward. However, this abdication of aesthetics is in itself an aesthetic, though a different kind, less conscious, more primal -- which is perhaps its appeal. The greatest aesthetic triumph of the geo-tube is the manufactured Amwaj Island in Bahrain, beneath the sands of which lie the world's largest geo-tube installation.[10] And the way that the geo-tube is unceremoniously dumped into streams on Filipino golf courses and along Mexican beaches and on the coastal dune-scape of Virginian spaceports captures exactly the unconscious nature of this un-aesthetic, and the weirdness of the geo-tube.

Aerial view of Amwaj Island, Bahrain. As described by Geotec Associates, "The key element in the successful design, construction and completion of the Amwaj Island development project was the use of sand filled Geotubes to form the island perimeter for containment of 12 million cubic meters of dredged sand that formed the basic platform for the development project. The island perimeter was constructed by hydraulically filling Geotubes with local sands from proposed navigation channels and marinas serving residents and businesses on the main island of Bahrain...Amwaj Island development will provide a variety of amenities such as living in beach front properties, hotels, restaurants, recreation parks, theaters, marinas, and golf courses."[11]

The Anthropocene may have begun accidentally and unnoticed, but now that we have seen it, it is necessarily propelled by choice, by our awareness of our capacity to shape it.[12] As a working material, geo-tubes suggest the possibility of a new soft architecture that at once works with and against geologic forces.[13] In a very physical and material way, these horizontal monoliths open the door for architecture as process, as the conditions under which they are deployed cannot help but suggest structures in the process of accumulating and structures in the process of dissolving.

Geo-tubes embody a new vernacular of engineered geology, of infrastructures for the self-aware Anthropocene. They surprise and captivate us, just as elevated freeway interchanges and massive dams captured the cultural imagination of the last century.[14] They are a harbinger of things to come. ∎

---

10 Per conversations by the authors with representatives of TenCate, 1 September 2011.

11 Fowler, Jack, et al. "Amwaj Island Constructed with Geotubes, Bahrain," *Geotec Associates*: http://www.geotec.biz/publications/Amwaj%20Islands%20Constucted%20with%20Geotubes.pdf.

12 For a longer exposition of a similar argument, see Andrew Revkin, "Embracing the Anthropocene," *The New York Times* [Dot Earth weblog], 20 May 2011: http://dotearth.blogs. nytimes.com/2011/05/20/embracing-the-anthropocene/.

13 In a piece, "Dredge," for the forthcoming publication *Bracket* 2 (published by Actar and edited by Lola Sheppard and Neeraj Bhatia), we further explore this link between "soft" architecture and the Dredge Cycle.

14 See David E. Nye, *American Technological Sublime* (Cambridge: MIT Press, 1994).

Christian Neal MilNeil

# 10. INNER-CITY GLACIERS

Inner-city "snow dumps" are holding areas for snow collected from city streets during the winter months. They relocate, consolidate, and reveal a condensed record of a city's air and water pollution. Image courtesy of Christian Neal MilNeil.

There is a sediment that hangs in a haze above city streets: a low-dose toxic dust of lead and chromates from tire wear, clouds of carbon soot mixed with hydrocarbon gases and fine particles of nitrates, sulfates, and other metals from exhaust pipes. At every stoplight, worn brake linings leave behind microscopic flakes of copper, zinc, and lead.

These automotive *disjecta membra* are rarely visible; you might find them as a dirty streak on a handkerchief after wiping your brow on a humid day, or in the accumulated dust of an a/c unit's intake. In most situations, these particles either disperse in the winds, or wash away into storm drains during rain showers.

But for a few months each year in cities of the upper latitudes, when precipitation falls as snow and remains above-ground in snowbanks for weeks at a time, these sediments have a chance to accumulate in impressive and highly-visible roadside deposits. Many of these cities collect and stockpile their snow in massive municipal "snow dumps" in order to keep city streets and sidewalks clear, thus consolidating the roadside's silt even further.

When city crews collect snow and pile it in snow dumps, they scour microscopic material first from the atmosphere, then from the ground, just as a glacier would. When the snow and ice finally melts at the onset of summer, these accumulated materials leave behind a terminal moraine of filth, typically one or two feet deep, that can serve as a sedimentary record of the city's winter. Image courtesy of Christian Neal MilNeil.

Abandoned or little-used industrial sites seem to be popular sites for snow dumps: Toronto uses a semi-abandoned quay in the Port Lands just outside of downtown; Elmhurst, Illinois dumps its snow into a deserted quarry. My hometown of Portland, Maine, dumps most of its snow in the empty flight path of the city airport, but an auxiliary snow dump gets built every winter just a few blocks from my home, in the Bayside neighborhood. It's an empty city-owned lot on the backside of downtown that had been a rail yard until the 1980s, and has been the hoped-for site of a few aborted redevelopment proposals since then. While it waits for high-rise office buildings, the city instead builds there a high-rise pile of snow, which the neighborhood half-jokingly calls "the Bayside Glacier."

Only half-jokingly, though, because there are a few legitimate reasons why it's apt to think of urban snow dumps as glaciers. First, there's their spectacular bulk and durability: the Bayside glacier can grow to a height of several stories, rising above the industrial neighborhood's low-slung warehouses to compete on the skyline with Portland's modest collection of downtown office buildings. The snow and ice inside frequently remains well into June, insulated from the sun under a protective mantle of accumulated road grime.

And just like a glacier, snow dumps are a legitimate geological force in the city's landscape. As snow falls from the sky, it scours out the microscopic pollutants of the city's atmosphere and carries it to the ground. Snow that hits the pavement of city streets and sidewalks also binds to road salt, oil, and antifreeze. All this gets plowed to the curbside, where the snowbanks collect larger items like forgotten bicycles, garbage bags, dog shit, piles of leaves, and other street detritus. In the days following a snowstorm, convoys of dump trucks will crawl along the streets behind a huge municipal snow-blower, which consumes entire snowbanks in its giant spinning blades and regurgitates them into the waiting trucks.

Thus the city's road grime and curbside filth, all of which were swept away weekly during the warmer months by street sweepers or the occasional rainstorm, instead get bound up in ice and concentrated in the city's glaciers for weeks at a time.

In late March and April, the top layers of the glaciers begin to melt away, revealing a sedimentary record of the winter. Near the bottom, a layer of leaves indicates the November storm that struck before the city crews had a chance to haul away the autumn's leaves; further up, a thicker-than-usual streak of garbage might indicate a storm that coincided with trash pickup day. In a process that glaciologists call preferential elution, microscopic impurities migrate to the surface while the outer layers of snow and ice melt or sublimate into the air, such that the entire thing resembles less and less a pile of snow, and more and more a pile of muddy filth.

When the last of the snow and ice finally melts at the onset of summer, these accumulated materials leave behind a terminal moraine of toxic mud, typically one or two feet deep—a visible distillation of a winter's worth of city air.

And so for the balance of the year, on an empty lot on the edge of downtown, two acres of gravelly filth remind us of what it is we filter through our lungs, every day, year-round—a spectacle even more humbling than a five-story pile of snow. ∎

While snow dumps are typically considered a blight, their ability to consolidate pollution over the course of an entire season, and make this collection visible with the onset of spring, gives them value to anyone who seeks to see what we typically think of as invisible. Image courtesy of Christian Neal MilNeil.

"We know when we build a bridge it will not last."
— representative from the Icelandic Road Administration, May 5, 2012

aerial view of Iceland's Ring Road
image courtesy Janike Kampevold Larsen 2012

Janike Kampevold Larsen

# 11. IMAGINING THE GEOLOGIC

In *The Tertiary History of the Grand Cañon District* (1882), US surveyor Clarence E. Dutton writes: "Great innovations, whether in art, literature, in science or in nature, seldom take the world by storm. They must be understood before they can be estimated, and must be cultivated before they can be understood."[1]

Dutton himself cultivates his experience of the Grand Canyon by making imagination a tool for fascination. He describes the impression of geologic forces upon himself. In order to explain how he perceives the magnificence of the rock faces he sets them in motion to encourage an understanding both of their making and their aesthetic impression: "the Vermillion Cliffs send off buttes," "the entablature [...] breaks into lofty truncated towers," "the towers of Short Creek burst into view" (53, 54). The vocabulary is architectural. The canyon is abundant with columns, buttes, terraces, promontories, bays, towers, churches, temples, theaters, and avenues: "As we move outwards towards the center of the grand avenue the immensity and beautiful proportions of the walls develop" (87).

The impression of movement seems first and foremost to come from the geologic masses' own agency. The canyon is a plastic scenario of ripping forces and flowing substances: "The [...] escarpments stretch their courses in every direction, here fronting towards us, there averted; now receding behind a nearer mass, and again emerging from an unseen alcove" (60). When Dutton describes a wall as "stretching athwart the line of vision" (86), he describes not the agency of the phenomenological gaze, but a geology that blocks that gaze's access. Dutton's animated perception displays the canyon as an active agent, as something showing itself to him. The canyon stages an unprecedented meeting between man and geologic forces and presence at unimaginable scales. Dutton is being confronted with earth's time.

*Tertiary History of the Grand Cañon District* fully demonstrates how knowledge and recognition of geological processes may enrich the experience of a particular tract of land, in this case the "veritable wonderland" (36) that is the Grand Canyon, much like Eric W. Sanderson's *Mannahatta* (2009) and smudge studio's *Geologic City* (2011) add to New York, activating an imagining of its past state and human making, and not the least, activating an imagining of its future.[2]

---

1 Clarence E. Dutton, *Tertiary History of the Grand Cañon District* (Tucson: The University of Arizona Press, 2001), 142. First published by the U.S. Geographical Survey (Washington, DC: 1882). Hereafter referred to by page number parenthetically throughout essay.

2 In *Mannahatta: A Natural History of New York City*, Eric W. Sanderson and his team of scientists reconstruct Manhattan of 1609 and thus add a layer to present day geography and the experience of the city. In *Geologic City*, smudge studio brings our attention to urban "events" featuring geologic matter and significance, brownstown buildings, scaffolding, Manhattanhenge and the city's closeness to the "canyon" in its waters.

By "leaps of the mind" and "slight effort[s] of the imagination" Dutton cultivates his sights in order to render them conceivable. As part of a detailed procedure for visualizing a reconstruction of the Grand Canyon's smooth Carboniferous platform, Dutton performs an imaginary depression of the freshwater beds of Eocene age, which are now located between 10,000 and 11,000 feet above sea level: "To find their position at the beginning of Tertiary time we must in imagination depress them that amount" (71). Imagination also carries Dutton to unforeseen terrains: "Could the imagination blanch those colors, it might compare them with vast icebergs, rent from the front of a glacier and floating majestically out to sea, only here it is the parent mass that recedes, melting away through the ages, while its offspring stands still." (54).

Robert Smithson performs a very similar exercise of setting geologic material in motion by help of imagination in his essay on Frederic Law Olmsted and Central Park:

> Imagine yourself in Central Park one million years ago. You would be standing on a vast ice sheet, a 4,000-mile glacial wall, as much as 2,000 feet thick. Alone on the vast glacier, you would not sense its slow crushing, scraping, ripping movement as it advanced south leaving great masses of rock debris in its wake. Under the frozen depths, where the carousel now stands, you would not notice the effect on the bedrock as the glacier dragged itself along.[3]

The "slight effort of the imagination" employed by Dutton and Smithson, helps illuminate that almost inexplicable fascination we feel when looking at exposures of geologic stratification revealed by present day engineering and construction—such as road cuts and surface mines. One could say that fascination with elemental and geologic presence in our environment is a fascination in the face of what is not us, what is not human, and which has its own agency. For French philosopher Jean-Luc Nancy, it is a fascination with something that is *not* landscape.

To Nancy, that which escapes the category of landscape also escapes us. The ground of the landscape, its base materiality, is always mediated by landscape as a cultural construct. We have no access to the ground as unmediated. To explain this, Nancy compares the relationship between "landscape" and "land," to how the image stands in a relation to that which it is an image *of*. The image is an image of the world, of the world it is made in and arises from. Once pictured however, this world, or ground if you like, disappears in the image. This is simpler than it sounds: the image lets go of what it portrays by portraying it! The image is something else, it has changed what it saw and it occupies place with the same authority.

Nancy further launches the idea of country (*pays*) as something that is not a politically, nationally, culturally and socially formed imaginary construct. It is a corner of land (of the earth) that one might belong to. Landscape (as image), on the other hand, is something that distances us from the country. As Christianity took its hold on landscape and substituted the noisy pagan gods living in the landscape with the ephemeral true God, it expelled all presence from it. The modern era provides no immediate relation to land, no unmediated reciprocity, compared to what the pagan gods offered by being a live part of the landscape. And this establishes a distance between the land and the culturally produced onlooker. To the extent that we are wrapped up in a notion of landscapes as visual and perspectivized scenarios, we are missing a sense of the world as an abundance of material without meaning, except for the differences that exist between shapes and forms, between tree and rock. Nancy opens up the possibility for a *sense* of the empty, meaningless landscape that the pagan gods surrendered to the Christian God, of land as nothing more or less than mass and presence. When the

---

3  Robert Smithson, "Fredrick Law Olmsted and the Dialectic Landscape (1973)," in Robert Smithson, *The Collected Writings*, ed. Jack Flam (Berkeley: University of California Press, 1996), 157.

distance of culturally produced meanings and perspectives is absent, we sense the ground in its desolateness, its lack of meaning and of gods: "there is no presence, there is no access to an 'elsewhere' that is not itself 'here.'"[4]

The American sublime, as suggested by David E. Nye, lies in novelty, and corresponds with Dutton's impression of sublimity while facing the Grand Canyon's "forms so new to the culture of civilized races" (90). For Dutton, the sublime lies exactly in encountering forms not yet part of the human categories for understanding: "In this far-off desert are forms which surprise us by their unaccustomed character. We find at first no place for them in the range of our conventional notions" (90-91).

Today, it is through and by technological and mechanical work that the earth is revealed. But the novelty for us, now, does not consist in the mechanized revelation of its unseen forms. It consists in the formlessness of the matter revealed, as well as in the deep time that it prompts us to imagine. How might we, expanding on Dutton's imaginary animation of geologic form and, by way of Smithson's imaginary animation of Central Park, move on to a conception of the "geologic now" as a new mode of awareness?

At the Norwegian Road Museum lies an old photo album. It was found in the office of the famous road director Hans H. Krag, the Norwegian equivalent to the obsessed "master builder" of New York, Robert Moses, only lacking the latter's more outrageous sides. Krag was responsible for most of the Norwegian highways built 1880-1903 in which time span the gross part of the country's mountain pass roads were built. His album of photographs shows planned roads, indicated by a red line drawn straight across a vertical slab running into a fjord, newly finished roads, and exotic road engineering forms, like half tunnels and glacial tunnels. The photos do of course most of all display the fascination with engineering possibilities, the marvel of human made forms in a nature not easily conquered. They also however testify to the marvel of geology released in the road building practices of the late 19th century.

Road construction Saude-Storkjær. Image from Road Director Hans H. Krag's foto album, 1900. Image courtesy of the Norwegian Road Museum.

4  Jean-Luc Nancy, *The Ground of the Image*, trans. Jeff Fort, Perspectives in Continental Philosophy (New York: Fordham University Press, 2005), 59.

Two photos stand out as something more than documentation of conquest, and they are astonishingly similar. Both are shot from inside of newly blasted tunnels, roads not finished. In both photos there are figures standing inside the tunnel gazing out of holes through which light flows in. The figures have an air of contemplation. These are not just construction scenarios. Rather they display an artistic wonder in the face of a new landscape or landscape situation, that of looking at the landscape from inside of its matter. The road construction site has formed a space in which nature seeps in and is displayed: an inverted museum.

And still the photos show an attention to the geologic presence of the inside of the mountain. The unknown photographer's sense for material form, light effects and scale recalls romantic painting from a century before, like those of Caspar Wolff's "Eine Jura-Höhle" from 1773, or Joseph Wright of Derby's "Cavern in the Gulf of Salerno," 1774, or his "A Cavern, Morning" from the same year.

Caspar Wolff: Eine Jura-Höhle, 1773.

Road cuttings are human made structures that allow us to co exist with a legible trace of deep time as rock faces, stratification, old riverbanks and just plain dirt is shoveled up and thus displayed. However, they also project into the deep future, Steven Ian Dutch claims in "Geosphere, The Earth has a Future." Structures likely to exist in a million years do not contain steel (as it would corrode). Instead, they include "open-pit mines, earth-fill flank dams, and large landfills." He also adds "large highway and canal cuts."[5]

To enjoy road cuts however, one must not only have a sense of geologic presence, but for the ordinary, for repetition of views, and not the least for that which we pass at great speeds.

---

5  Steven Ian Dutch, "The Earth Has a Future," *Geosphere* 2.3 (2006): 113–124; doi:10.1130/GES00012.1.

Norwegian artists Ingrid Book & Carina Hedén displayed their series *Produced Landscapes* at Oslo School of Architecture and Design in 2009. For their exploration of roadside landscapes they have chosen the least spectacular stretch of motorway in Scandinavia, the E6 between Oslo and Gothenburg. The road cuts through a semi-industrial cultural landscape broken by shorter stretches of grove—and cuttings. The rock faces are marked by industrial process; the slabs are brutally cracked and bear marks of drilling and blasting. Book & Hedén's 30 photographs catch the generic landscape of the motorway cuttings, the typicality of the Scandinavian gneiss and granites. However, as always in series based on typology, this series manages to display the untypical of each cutting: colors, geologic composition, and rock qualities vary in the tableaux revealed by the photographer. The project thus achieves some of the same effects as some of the typology studies in Gerhard Richter's *Atlas*.[6] His series of slightly shifted views of gardens, rubble, shrubs, and mountain views unfold the immense variety produced by a slight shift of perspective, and thus reflect a deep desire to say something of how ordinary, often messy, landscapes affect us, and how we are drawn to them. Book & Hedén ask the same question: why are we attracted to haphazardly blasted roads cuts?

The photographic closeness to the rocks produces surprising effects as one projects one's imagination onto their materiality. One of the cuttings looks soft, like draped velvet flowing down the slope. Another one looks like a cut-out part of El Capitan, wide red slabs of sandy granite, yet another a roadside "mesa" displayed to northern hemispheres. Still others are framed by dirt and stone masses. The decorative grasses so typical of roadsides in Europe have not yet been planted and the sites thus retain a construction quality. The ground is exposed in a more raw way and thereby the sites escape the leveled character that Uvedale Price shunned as it concealed "the ground itself."[7]

The exposed rock of cuttings like these, or by the one on the Canadian Shield shot by Jamie Kruse, exhibit to us a time that we have no access to: the slow time of geology. The temporality of the rock has been transgressed in and by the blasting of it. As happens in the Grand

6  Specifically in the series' *Hahnwald* 2006 and *Sils Maria* 2006; see Gerhard Richter, *Atlas* (Köln: Verlag der Buchhandlung Walther König, 2006), 757-783.

7  Uvedale Price, *Essays on the Picturesque, As Compared with the Sublime and the Beautiful: And, On the Use of Studying Pictures, for the Purpose of Improving Real Landscape* (London: J. Mawman, 1810).

Canyon, the rock becomes legible as its layers, patterns, and composition is revealed in the cutting. The cut, or incision in the rock, which we will not see clearly at speeds from the car, but all the more clearer in the camera's still frame, defies the extreme slowness and gravity of the rock. Thus the road cut displays an encounter between the slow geological process and a rapid perception of the material—one could say between the 19th century's fascination for geology as science and technique and our time's dependency on high speeds. They allow us access to a temporality and a materiality that our common speeds tear us away from. Ironically it is our infrastructure that allows us both.

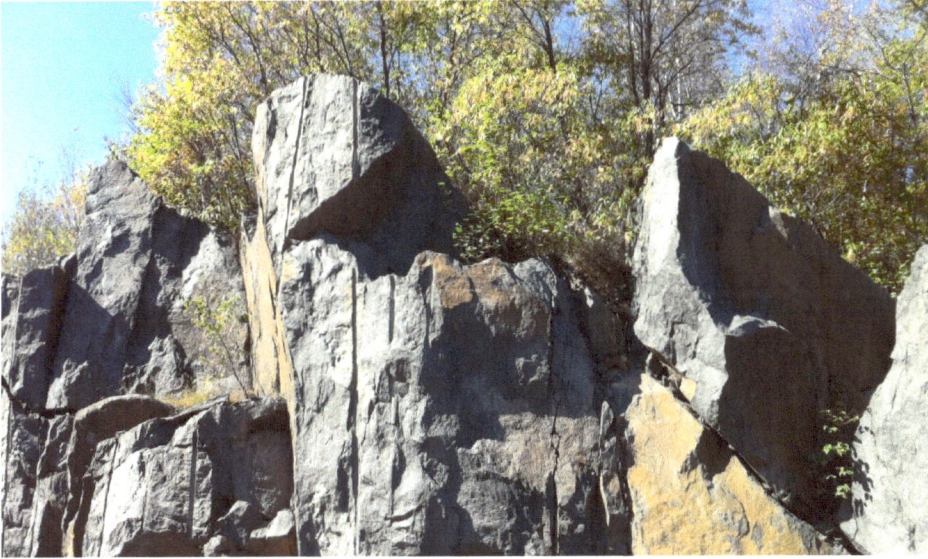

Road cut at the Canadian Shield. Image courtesy Jamie Kruse.

In the post *Digging Vermont. What lies beneath the green mountain state*, The Center for Land Use Interpretation observes the mountains of slate pieces formed by excavation residue from mines in the territories of Vermont. They recognize something that we might call an implicit fascination: "[R]emarkably few seem to mind how it looks." In conclusion to this post CLUI proceeds to deliver what is, for them, a rare statement of aesthetic evaluation:

> Maybe we will decide, though, that terrestrial excavation's honest depiction of our collective consumption can be alluring, or even beautiful, like the picturesque old quarries that litter the state already. Maybe one day the extraction of the mineral resources of the Green Mountain State will be more highly appreciated, as unlike ski condos, it operates, for the most part, below grade, in the gray area under the green.[8]

Here, in one of quite few places where CLUI actually offers an aesthetic evaluation of the many landscape typologies they document, they give expression to the need for a process of habituation not altogether different from Clarence E. Dutton's credo. The cultivation of contemporary landscape typologies like those produced by ground activities of Vermont, contributes to develop and uncover strategies to re-conceptualize our landscape. They also display the necessity of imagination in articulating the significance of contemporary land-scapes, as does Book & Hedén's images. This does not, and maybe should not imply investing them with meaning, or even assigning them specific uses. It could mean simply recognizing their agency to produce an effect in us by their mere presence. ∎

8  CLUI, "Digging Vermont: What Lies Beneath the Green Mountain State," *The Lay of the Land Newsletter*, Winter 2011, http://www.clui.org/lotl/v34/digging_vermont.html

Oliver Goodhall & David Benqué

## 12. ULTRA-DIAMOND / SUPER-VALUE[1]

The diamond, as we know it, is a unique example of marketing and monetization of a geological resource. Through careful supply-control, advertising, and cultural massaging, the industry has managed to mythologize a geological material in cultures around the world. Throughout the 20th century, rituals, expectation, and meaning have been intentionally crafted around the diamond through tales of rarity and carat-value. A complete mythology, ranging from sparkling wedding rings to shady deals and overworked mines, surrounds each of these stones and furthers the mystery as well as desire.

As a counterpoint to this aesthetic industry, a new functional aspect is emerging as diamonds are grown in the lab with ever increasing control and huge promises for technological applications. The unique properties of diamond as a super-material open up potentially revolutionary breakthroughs in fields as varied as quantum computing, electronics, biosensors, and clean energy.

As lab-grown diamonds become more recognised and enable further progress, will they also achieve a new cultural status? Until now, the aesthetic and the technological are carefully kept separate to preserve market value and cultural narratives. We can easily imagine that gemstone dealers have no interest in promoting the fact that chemically perfect diamonds are now routinely grown in laboratories. Synthetic diamonds have become recognised for enabling further technological advances and while doing so achieve a new cultural status to rival their natural counterparts

As synthetics become more important and celebrated as agents of our technological progress, how will this impact their place in society? And, functionally, is the eternal promise of diamond about to deliver very tangible results?

*All artefacts, images, and texts presented in this essay are purely fictional. Any resemblance to real persons, situtions, or products is purely coincidental.*

**Natural Synthetic**

|  | Natural | Synthetic |
|---|---|---|
|  | love | science |
| fig 1. | eternity | singularity |
| Comparative table for | priceless | price point |
| the aspiring synthetic | endless | limitless |
| diamond agency | enduring | reliable |
| copy-writer | promise | promise |
|  | earth's core | lab gown |
|  | flawless | perfect |
|  | forever | infinite |
|  | more desired | more powerful |
|  | romantic | technological |
|  | I do | can do |
|  | finely cut | finely fabricated |
|  | dream today | tomorrow's dream's |
|  | expectation | anticipation |
|  | occasion | progress |
|  | social status | mateial status |
|  | DeBeers | Element 6 |

---

1 A visual essay that presents fictional devices that enable the celebration, transport and valuation of diamond super-materials. These devices are staged as the supporting characters in the true, currently unfolding story of man-made geology.

## #053 Attaché-Case
Patent 3492834

Abstract title:
Expandable, divided attaché case to support Chemical Vapour Deposition diamond wafers (<0.2mm thk) securely and discreetly masking dimensions.

Description:
An attaché case with an outer enclosing case and a protective insert within to hold specifically engineered materials. The inner case may be visually impregnable. A scalable clip system secures engineered wafers within separate dividers numbering up to 20, each pivoting on secure fixing along a single edge. [*Clipping mechanism and secure handle subject to separate cited patent documents.*]

**#072 Drill Bit Replica**
Policy #3290098

Value Assessment for Item #0387
Category: Fine Art and Jewellery

Described for valuation is a unique, one-off piece, part of the Oppenheimer family's private collection. The item is believed to be a decorative replica of an industrial diamond coated drilling bit, used for oil drilling. The origin of the item is unclear, and there is no manufacturer's stamp to be found.

The item is believed to have been received as a gift to the late P. Oppenheimer as part of an informal transaction.

*Cast formed 22K rose "crown" gold alloy body with 18K gold details and 408 0.03 carat micro-machined D-VVS1 diamonds*
*Estimated Value: circa $300,000*

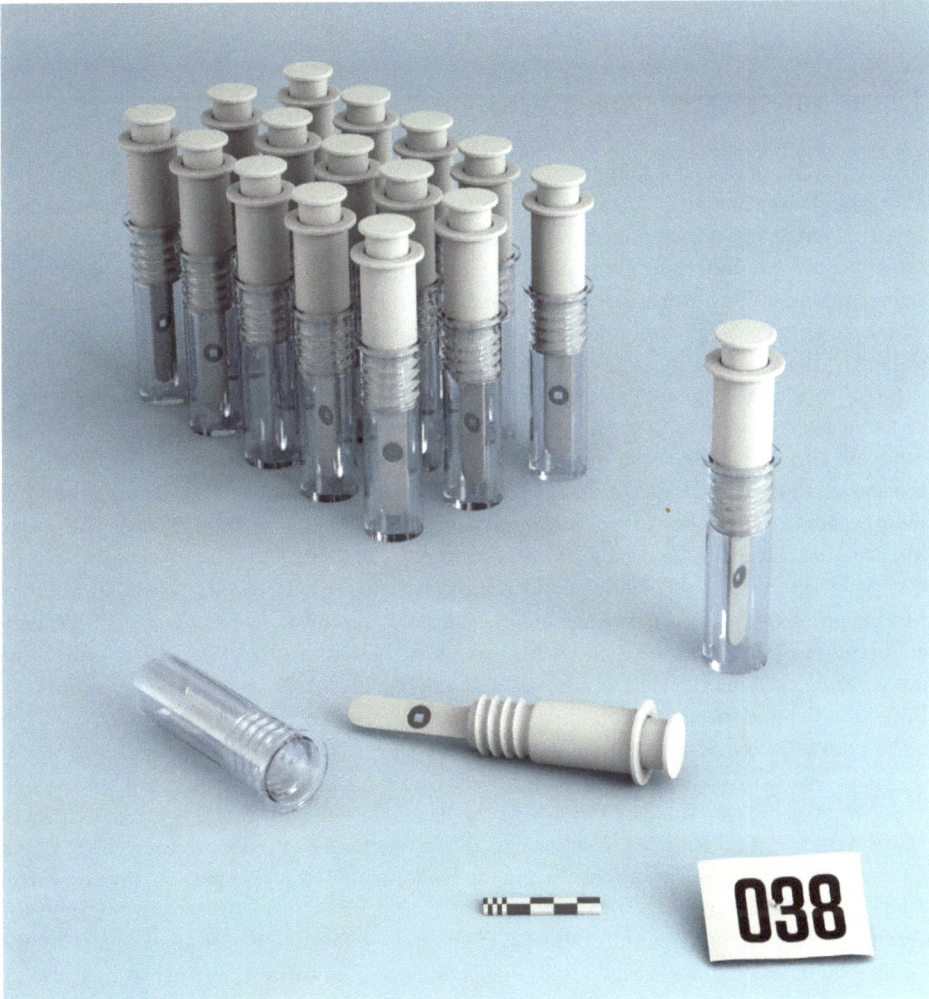

## #038 Boron Test
Instructions: Testing for traces of Boron doping & superconductivity growth modification

1. Swab or enclose sample of object to be tested.
2. Screw swabbing device into the clear plastic  tube until airtight.
3. Pull plunger upwards to break seal.
4. Allow air to circulate for minimum 30 seconds.
5. Check swatch indicator and refer to colour chart on reverse of packet.
6. Red = traces of doping superconductor level (check spectrum chart).
7. Blue = confirmation of negative reading.

This test checks for Boron traces and particles assuming typical $CH_4$ mixing ratio ~1-2vol%. Test has been lab tested to accuracy of 95%.

*The manufacturers of any part of this equipment take no liability for the results under field conditions.* ■

Julia Kagan

## 13. THE NOT-SO-SOLID EARTH: REMEMBERING NEW MADRID

"What would happen to our thinking about nature if we experienced [non-human] materialities as actants?" Jane Bennett asks in *Vibrant Matter: A Political Ecology of Things*. One result might be a different way of thinking about earthquakes. The stresses and fault lines and tectonic-plate collisions that produce them are material forces whose actions can reshape the world in ways that profoundly affect human life. Earthquakes remind us that human lives take place against a background of geologic time and that human agency, however nature-bending, is ultimately shaped by the non-human context in which it operates.

Living in a known quake zone where Earth's tectonic plates meet means inhabiting a geology that feels very much alive, not buried in the remote past. Think Japan, Chile, and the other nations along the Pacific "rim of fire," not to mention its eastern border: California, the American northwest and Alaska. This doesn't mean that those who reside far from plate margins live in safety. Perhaps the strongest earthquakes to hit the United States in recorded history were intraplate, as geologists term it, not at a tectonic edge. They happened exactly 200 years ago in the middle of the North American Plate, in a place called New Madrid in the Mississippi River Valley. Three huge quakes hit from December to February 1811-12, with estimated magnitudes (the seismograph was not invented until 1896) as high as 8.1. The anniversary is a reminder that reverberations from immeasurably ancient geologic events can burst into the present without warning.

Americans who live on the East Coast got their own intraplate reality check on August 23, 2011. As many were returning to work after lunch, a 5.8 magnitude quake struck Mineral, Virginia. Because the Earth's crust in the East is brittle and relatively unbroken, shock waves were felt as far south as Alabama and all the way to Canada. (By contrast, the frequent quakes of the West Coast result in chopped-up rock that deflects wave motion, making earthquake impact much more local.) Washington, DC, shook hard enough to open cracks in the Washington Monument; authorities closed the site "indefinitely" pending repair. At the National Cathedral, a stone angel and other pieces of the central tower fell to the ground. New York City evacuated courthouses and many office buildings.

Although the eastern U.S. borders the Atlantic Ocean, it's not at the edge of the North American Plate. The plate's eastern border is the Mid-Atlantic Ridge at the center of the ocean, surfacing only at Iceland. Intraplate quakes originate in geologic weaknesses within a plate that succumb to stressors as pieces of the Earth shift over time. The Virginia temblor occurred near ancient faults involved in creating the Adirondack Mountains some 250 million years ago, according to Columbia University's Lamont-Doherty Earth Observatory. "This is a wake-up call to take earthquake preparedness in the northeast more seriously," said seismologist Arthur Lerner-Lam, its interim director. "Let's not turn over and go back to sleep."

The danger may be even greater in the New Madrid Seismic Zone (NMSZ), which sits over historically active faults within the rift valley that formed more than 500 million years ago in the crust under southeastern Missouri, northeastern Arkansas, and northwestern Tennessee. Later, this Reelfoot Rift system filled in with a thick layer of sediment. Eventually, the Mississippi River and its valley formed over it. By 1811, the land was mostly wilderness and wetland, with a few small settlements along the river, of which New Madrid, now a small town in Missouri, was the largest. Its population was about 400, according to Jay Feldman's *When the Mississippi Ran Backwards*, a history of the disaster.

The first of three immense quakes struck on December 16th, 1811, at 2:15 a.m. It was a warm December night when at least some of the townspeople were recovering from a Sunday night dance. It began with loud rumbling, then sleepers were jolted awake as furniture flew across the room. People ran into the street in their nightclothes. As the earth cracked and the river roiled, plumes of sulfur-smelling vapor filled the air along with the cries, not just of humans, but of farm animals and wild birds. Trees swayed and snapped, chimneys fell. Giant cracks appeared; in the nearby town of Little Prairie (now Carruthersville, Missouri), a granary and smokehouse slid into a fissure, which then closed, swallowing them. It felt like the end of the world. "The loud, hoarse roaring which attended the earthquake, together with the cries, screams and yells of the people seem still roaring in my ears," one New Madrid resident wrote a friend.

Two other temblors, also immense, followed on January 23rd and February 7th of 1812. In addition to the big quakes there were dozens, if not hundreds, of smaller shocks. Houses fell, the ground liquefied, the Mississippi ran backwards and a sizable lake was born, all just 165 kilometers north of Memphis. Geo met bio with a violence that still surprises.

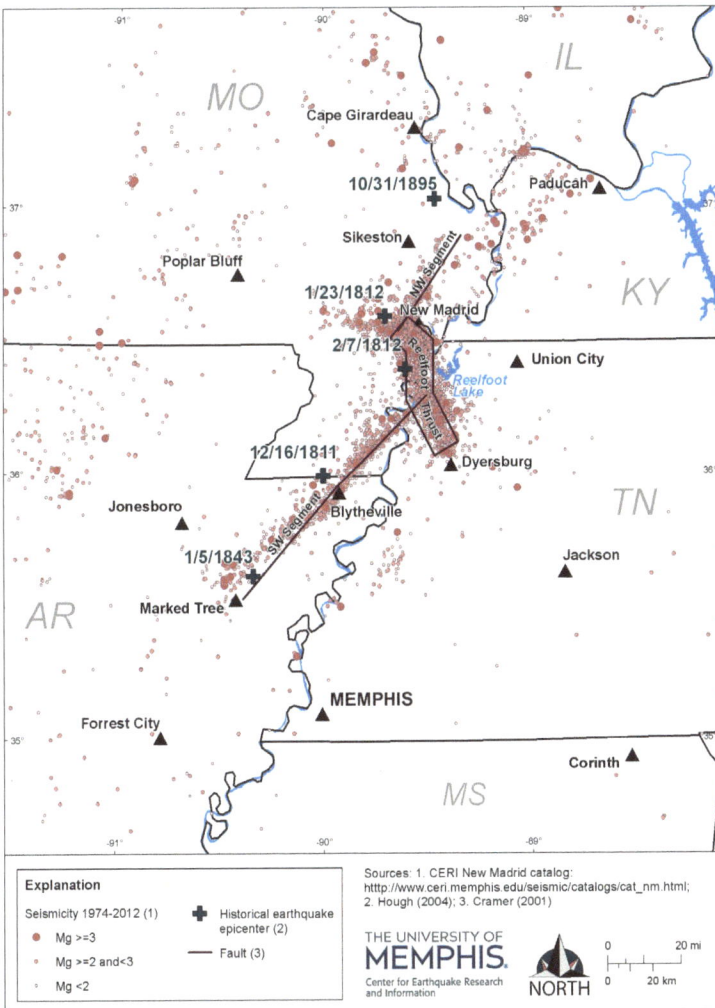

Figure 1: Seismic Map. Courtesy of University of Memphis: Center for Earthquake Research and Information

To this day, the New Madrid region "remains the most seismically active area east of the Rocky Mountains" according to the Seismological Society of America (SSA), which marked the anniversary by holding its 2011 annual meeting in Memphis, followed by a field trip to the geological remnants, which this reporter attended. The SSA calls the buried rift valley [see Figure 1] "a serious threat to the metropolitan and agricultural areas of the central United States and to the numerous lifelines that pass through [it]." These include four of the nation's five major natural gas pipelines. Since 1812, there have been at least 28 damaging earthquakes, with estimated moment magnitudes between 4.2 and 6.4. Meantime, paleoseismologists have found evidence of previous New Madrids in 300, 900 and 1450 C.E.; some believe earlier ones struck in 1000 and 2350 B.C.E.

The most visible sign of the destruction is a pattern of sand blows, stretches of sand that boiled up from below when quakes liquefied the earth. These pale patches still stand out clearly against the darker Mississippi mud. The SSA group's guide, paleoseismologist Martitia Tuttle, whose research established many of these dates, showed us examples in the Yarbro excavation near Blytheville, Arkansas. Figure 2 shows an upper sand blow and a lower one, plus a sand dike, remains of the channel through which sand-bearing water erupted to the surface.

Figure 2: Martitia Tuttle, left, and Roy Van Arsdale at Pawpaw Creek. Photo by Julia Kagan.

Vivid eyewitness accounts recall some of these explosions. Thirty miles downriver from New Madrid, an eight-year-old in the town of Little Prairie saw the ground "rolling in waves of a few feet in height, with a visible depression between. By and by those swells burst, throwing up large volumes of water, sand and a species of charcoal," according to a report printed in *When the Mississippi Ran Backwards*.

At a later stop, Tuttle and geologist Roy Van Arsdale of the University of Memphis scrambled up the sheer 50-foot bluff over Pawpaw Creek to point out the layers of sand and rock left by New Madrid and other geologic events. As the rest of the group stood across the creek far below them, they showed [Figure 3] a deep base, a gravel layer that marks the original level of the Mississippi River, layers of Pleistocene rock, and much evidence of landslides. The Reelfoot River, into which the creek once emptied, disappeared after New Madrid.

Figure 3: Sand blow, Yarbro excavation. Photo by Julia Kagan.

Could it happen again? The Federal Emergency Management Association sees "potential for a catastrophic earthquake equivalent to those in the 1800s, centered on the New Madrid zone," as Mike Pawlowski, a FEMA instant response section chief, told a February 2006 hearing of the House Subcommittee on Economic Development, Public Buildings and Emergency Management. The probability: "about 10 percent in the next 50 years," according to Eugene Schweig of the U.S. Geological Survey. It might sound low to some people, but he says it's "actually quite high."

We could be getting close, at least by geological time standards: "For the last 2,000 years there have been sequences every 500 years," says U. S. Geological Survey seismologist Susan Hough. Author of three books on quakes, she measured the principal New Madrid quakes at about 7.0, lower than other estimates. "I think most scientists would say it's not prudent to assume it's over."

The emergency management community seems to be getting the message that geology isn't just a West Coast problem. On April 28, 2011, FEMA, the Central United States Earthquake Consortium (CUSEC), the U. S. Geological Survey, and the Institute for Business and Home Safety joined with state and local partners for the Great Central U.S. ShakeOut. More than three million people at nearly 10,000 sites signed up to take part in this multi-state earthquake drill for local communities in the NMSZ, modeled on exercises held in California.

A few weeks later FEMA, the Centers for Disease Control and Prevention, the Food and Drug Administration, and voluntary agencies such as the Red Cross—plus state emergency operations centers in Illinois, Indiana, Missouri, and Arkansas—participated in National

Level Exercise 2011. This dry run for handling a catastrophic 7.7 level earthquake on the New Madrid fault was "the first national exercise to look at a natural disaster," Jim Wilkinson, CUSEC's executive director, told the seismology meeting in a luncheon address on April 15th.

Though some activities were truncated by the need to respond to May 2011's real floods and tornadoes, more evidence of the power of non-human materialities, both projects took place. They evince a new openness on the part of government and business to the geologic underpinnings of our landscape and the need to respond to the suddenness with which these ancient forces can make themselves known.

It's interesting to juxtapose these developments with the concurrent realization that human activity is increasingly influencing geology. On the earthquake beat, debate rages about whether the deep underground wastewater disposal wells required by the controversial natural-gas extraction method called hydraulic fracking can trigger them. The Arkansas Oil and Gas Commission closed two disposal wells in July 2011 because of concern over the thousands of small temblors known as the Guy earthquake swarm, after the town where they appeared after drilling was underway. Similar links have been found to wastewater wells in Oklahoma and near Blackpool, England. U.S. Geological Survey scientists delivered a cautionary report about the relationship between oil and gas production and midcontinent quakes in the U.S. at the 2012 SSA meeting in San Diego. Then there's the growing evidence of a relationship between fossil fuel usage and global warming. Whatever the research finally tells us, geological consciousness is rising, as is the realization that not all the actionality on the world stage is human. ■

Jane Hutton

## 14. DISTRIBUTED EVIDENCE: MAPPING NAMED ERRATICS

This project maps a series of boulders that were plucked, transported, and deposited by the toe-line of the retreating Late-Wisconsin and pre-Wisconsin ice sheets in North America and subsequently named, relocated, modified, and celebrated by people. They are the glacially distributed sites of council meetings, picnics, political movements, and territorial markers. They are inscribed with discrepant personal, regional, and national narratives and at the same time they declare their foreign origin through their conspicuous mineral composition and form.

In the mid-19th century such boulders served as critical evidence for piecing together a theory of glaciation, and consequently an idea of geologic time and the location of humans within it. While their scientific importance waned in the 20th century, outside and along-side geology they are critical objects for reflections on time and space of various scales and consequences. They refer simultaneously to different moments in time: their geogenic forma-tion, their glacial deposition, and specific events in cultural history, as they are sometimes literally carved with a date. They are heavy, insistent markers in space, yet they indicate a remote origin—and therefore the journey between two sites—and they continue to be moved or changed by ensuing human forces. The naming, photographing, and feting of the boulders collated and mapped here signal their role as persistent devices for locating deep time within the present and grappling with the continuum of geological and human action.

At the Last Glacial Maximum, or Wisconsin Glacial Period in North America (26,000-20,000 years ago)[1], vast ice sheets extended across Northern Europe, Northwestern Asia, and much of the Andes. Acting as a mammoth material conveyor, flowing ice transported fragments of rock as it advanced. These blocks, bound tightly by the ice, scoured the surfaces that they passed over, abrading deep parallel grooves in the direction of ice flow. The majority of this rock material was deposited near to where it was picked up. A small fraction of it, usually that composed of harder minerals, travelled much further and was deposited, often on bedrock of a different mineralogical composition. German-Swiss geologist Jean de Charpentier described landscapes of the southern Jura Mountains that were conspicuously strewn with such boul-ders as *terrain erratique*.[2] The term originally coined to describe an anomalous landscape later came to describe the individual thing: the foreign boulder, the erratic.

Massive glacial erratics were found on German plains originating from Scandinavia, in Brandenburg originating from across the Baltic, and in St. Petersburg originating from some-where near Finland.[3] By the early 19th century, their distribution was considered "one of the most widespread, most striking and most inexplicable of geological phenomena." Alpine shepherds, closely observant of local landscape change over long durations, were the first to note the form of angular boulders downslope of retreating glaciers and convey their suspi-cions to early naturalist visitors. These localized data points, when aggregated, confirmed the speculation that existing glaciers had once been larger, and that ice, rather than the deluge, had distributed boulders throughout the valley. This led to the development of a theory of continental scale glaciation.

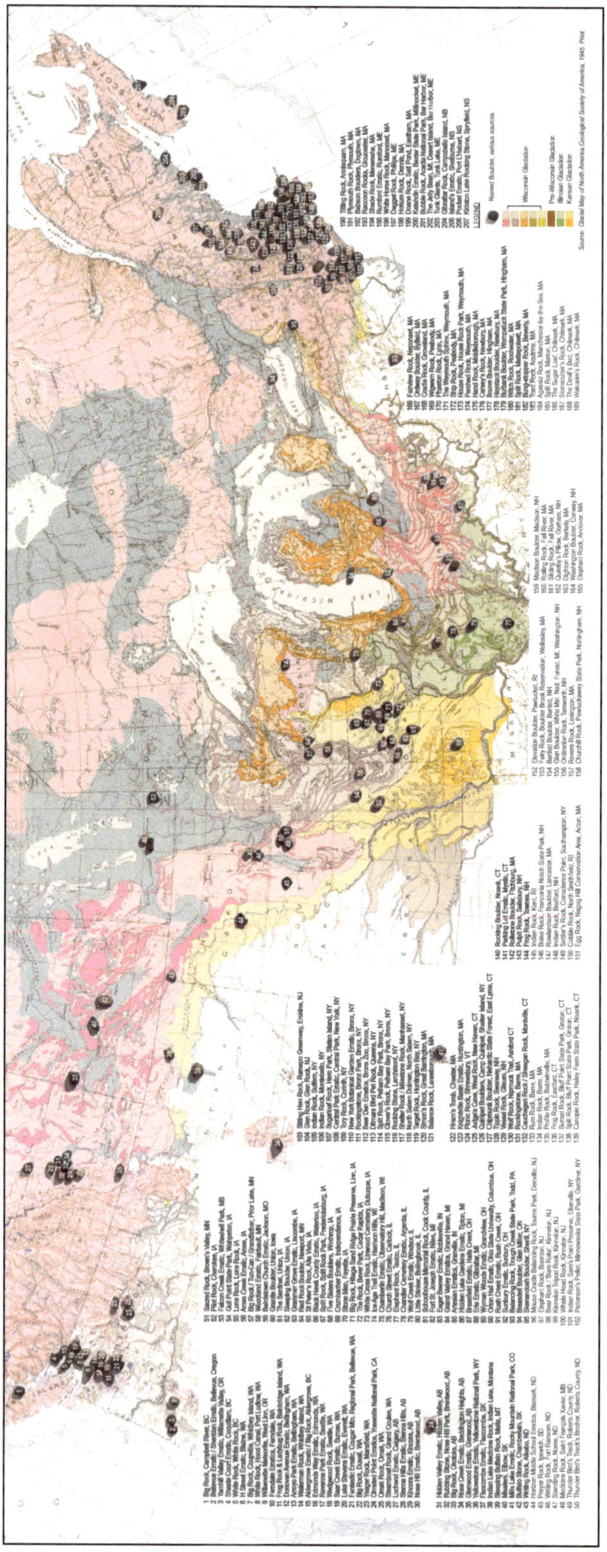

Figure 1: Map of Named Erratics. Visit the website: http://geologicnow.com/14_Hutton.php for a more detailed image.

The "Map of Named Erratics" (Fig. 1) likewise aggregates individual data points to draw a continental scale pattern of glaciation. In this case, the data points are some 207 culturally identified boulders distributed along the east to west glacial limit line. Named boulders were located from scientific and popular geological texts, town newspapers, local history museums, and for more recent documents, through geo-caching websites. Rocks were included as long as they had somehow been named, recorded, and characterized on an individual basis, and as a minimum criterion, whose name would be recognized by people who don't know each other. "An erratic on Mt. Washington" wouldn't make the cut, but "Sleeping Boulder" would. Rocks are often given the name of the river, town, or street that developed around it. Other times names describe the rock's likeness to an animal or object (Frog Rock, Whale Head Rock, Bung-Stopper Rock). A series of names describes the positioning of the rock, and its stability or instability (Tilting Rock [Fig.19], Balance Rock [Fig.8], Tripod Rock [Fig. 15]). There are five "Split Rocks," not surprisingly all of which are broken into pieces, indicating an interest in the idea of the boulder as a singular unit. There are seven boulders named "Indian Rock," which suggests the legacy of appropriating, stealing, and museum-izing of indigenous and often sacred boulder landmarks. The base drawing is a USGS map of glacial features (1945), which indicates the extents and surface deposits of the Wisconsin, Illinoian, and Kansan Ice Sheets, and then at a much finer grain indicates smaller scale glacial phenomena. A dotted branching symbol indicates the site of a boulder train; a small cross signifies areas where multiple erratics have been observed. Individual boulders, whose observation was instrumental in the understanding of these large-scale processes, however, slip out of the resolution of such maps.

(from left top to right bottom)

Fig. 2. Omak Rock, Omak, WA. Photo: Tim Greyhavens
Fig. 3. Ship Rock, Peabody, MA Photo Source: Gardner Collection of Photographs, Harvard College Library
Fig. 4. Rollstone Boulder, Fitchburg, MA Photo Source: Gardner Collection of Photographs, Harvard College Library

Fig. 5. Cauchegan Rock, Montville, CT Photo Source: Gardner Collection of Photographs, Harvard College Library
Fig. 6. Phaeton Rock, Lynn, MA Photo Source: Gardner Collection of Photographs, Harvard College Library
Fig. 7. Balance Rock, Lanesborough, MA Photo Source: Lanesborough Historical Commission
Fig. 8. Rollstone Boulder, Fitchburg, MA Photo Source: Peter Cristofono collection
Fig. 9. Kidstone Lake Rocking Stone, Kidstone Lake, Nova Scotia. Photo Source: Gardner Collection of Photographs, Harvard College Library
Fig. 10. Cobble Rock, North Smithfield, RI Photo Source: Peter Cristofono

(from left top to right bottom)

Fig. 11. Rum Rock, Barre, MA. Photo: Jim Moore
Fig. 12. Babson Boulder (Courage), Cape Ann, MA Photo: Jane Hutton
Fig. 13. White Rock, White Rock, British Columbia Photo: Kevin Korpi
Fig. 14. Bronx Zoo Rockingstone, NY Photo Source: Roswell Yates
Fig. 15. Tripod Rock, Kinnelon, NJ Photo: Crystalinks.com
Fig. 16. Madison Boulder, Madison, NY Photo Source: Gardner Collection of Photographs, Harvard College Library
Fig. 17. Prayer Rock, Ipswich, SD Photo: J. Stephen Conn
Fig. 18. Tilting Rock, Annisquam, MA Photo Source: Gardner Collection of Photographs, Harvard College Library
Fig. 19. Bubble Rock, Acadia National Park, Bar Harbor, ME Photo: Matt Lehrer

The photographs or rock portraits assembled here (Figs. 2-19) illustrate 17 of the mapped boulders and were taken from the 1880s to the present. The earliest photographs shown here were used by field geologists studying glacial evidence in New England. Other rock portraits range from promotional postcards to travel photographs from online photography databases. The photos depict ways that people measure, move, and inscribe the boulders, which can be understood as their attempts to think through their relation to geological forces. First, the typical rock portrait includes a scale figure (a person) standing next to, lying on top of, or pretending to support the weight of the boulder through a laboured pose (which is compulsory when visiting Bubble Rock, Acadia National Park, [Fig. 19]). The impossibility of physically moving these massive boulders, coupled with the knowledge that in fact they had been moved great distances, allows one to size up forces otherwise beyond comprehension. A second typical photo, and related to the first point, features the class of erratics that are perched on uneven ground and have the ability to "rock." These have long received extra attention. Both the Kidstone Lake Rocking Stone (Fig. 9) and the Bronx Zoo Rockingstone (Fig.14) are pictured as anticipating motion: at Kidstone Lake a man stands near the large lever installed to give the stone a boost, and in the Bronx Zoo portrait the stone is flanked by individuals who might give it a push. This ability to animate something so seemingly static opens up the possibility of it moving again. A third type of portrait features the inscription of dates and words as a means to commemorate recent time and meaning. These are seen from Plymouth Rock to the Babson Boulders (Fig. 12, "Courage"). Prayer Rock (Fig. 17), significant in Lakota accounts of the creation of the earth, was moved 68 miles to the front of the public library grounds in Ipswich, South Dakota. The rock is mentioned on a website listing Ipswich's major attractions, as is the mural that depicts the founding history of the town, in which the rock is prominently featured even as all traces of the Lakota have been erased. Prayer Rock is branded with a plaque that commemorates the erection of the library, not the rock, signifying the ways in which erratics are convenient media for manipulating time and history—the Lakota population present both at the time of colonial expansion and now, are relegated to a naturalized pre-history.

Gathering and resituating these individual anecdotes within the continental scale highlights the role of local, distributed observation as integral to the large-scale representation of geological processes. It also recognizes the ways humans grapple with time and spatial scales that are beyond comprehension. As scientific objects essential in the construction of a theory of deep time, and as popular artefacts to assist in contemplating and re-writing time, erratics link the seemingly unbridgeable chasm between human and geological scales. ∎

*–many thanks to Senta Burton for assistance with this project.*

1   Peter U. Clark, Arthur S. Dyke, Jeremy D. Shakun, Anders E. Carlson, Jorie Clark, Barbara Wohlfarth, Jerry X. Mitrovica, Steven W. Hostetler, and A. Marshall McCabe, "The Last Glacial Maximum," *American Association for the Advancement of Science*, 325.5941 (7 August 2009): 710 - 714.
2   Richard Foster Flint, *Glacial Geology And The Pleistocene Epoch* (New York: J. Wiley & Sons, 1947), 116.
3   Martin J. S. Rudwick, *Worlds Before Adam: The Reconstruction of Geohistory in the Age of Reform* (Chicago: University Of Chicago Press, 2008), 191.
4   Alexandre Brongniart, quoted in Rudwick, *Worlds Before Adam*, 186.
5   Ralph W. Dexter, "Historical Aspects of Agassiz's lectures on glacial geology (1860-61)," *Earth Sciences History* 8.1 (1989): 75-79.

Fukushima Nuclear Power Plant, © National Land Image Information (Color Aerial Photographs), Japan Ministry of Land, Infrastructure, Transport and Tourism, 1975.

"I am amazed by the fragile state of things despite all the economic and technological 'strength' Japan has been so proud of… Was our achievement of the past several decades a house of cards? The media often uses the phrase 'beyond assumption' for the disaster, meaning that its force was beyond architectural requirements. But I can't help sensing a more fundamental disruption between our norm and the reality."
— Toyo Ito, *Project Japan Metabolism Talks…*

Paul Lloyd Sargent

# 15. LANDSCAPES OF ERASURE: THE REMOVAL—AND PERSISTENCE—OF PLACE

## INTRODUCTION: HOW TO NAME A PLACE REMOVED?

What lexicon or neologisms might we use to speak of places removed by human force? Not by natural disaster, like the city of Pompeii or the Japanese fishing village of Minamisanriku, destroyed in the tsunami of 2011.[1] Nor by the horrors of war or catastrophes of technology, as in Virilio's obsession with the accident,[2] like those at Chernobyl and Three Mile Island, or the toxic morass at Love Canal in Niagara Falls.[3] Instead, I consider here what I am calling *landscapes of erasure*: the large-scale, intentional removal of socio-cultural place through processes of engineering, often due—but not limited—to industries of extraction, power generation, and shipping and transport. That is, I am exploring the erasure of material and socio-cultural landscape by human forces operating at the geologic scale because such scale reveals and magnifies the consequences of human interaction with the land, both intentional and otherwise.

---

1  Fuyubi Nakamura, "Memory in the Debris: The 3/11 Great Japan Earthquake and Tsunami," *Anthropology Today* 28.3 (2012): 20-23.
2  Paul Virilio, *The Original Accident* (Cambridge, UK: Polity, 2007).
3  See Lois Gibbs and Murray Levine, *Love Canal: My Story* (Albany: State University of New York Press, 2002) and Margaret Wooster, *Living Waters: Reading the Rivers of the Lower Great Lakes* (Albany: Excelsior Editions/ State University of New York Press, 2009).

Landscapes of Erasure: Lost Villages of Canada, Lock 21 at Dickinson's Landing under the St. Lawrence River, Long Sault, ON, May 2012.

## WORDS AND PLACES: TOPONYMS AND HYDRONYMS

"Local names—whether they belong to provinces, cities, and villages, or are the designations of rivers and mountains—are never mere arbitrary sounds, devoid of meaning. They may always be regarded as records of the past, inviting and rewarding a careful historical interpretation." Onomast and toponymist Rev. Isaac Taylor offered this thought on the legacies of place names in a 1864 essay entitled "Words and Places: Or Etymological Illustrations of History, Ethnology and Geography." Even in instances where "the original import of such names has faded away, or has become disguised in the lapse of ages," Taylor posits, a place name "may speak to us of events which written history has failed to commemorate."[4]

Taylor's text recounts geographic nomenclatures that dot the globe. He cites, for example, the *St. Lawrence River*, named for the Catholic feast of the 10th of August, when Jacques Cartier discovered the waterway, and implications in the toponym Wales, possibly from the German root *wal*, meaning "foreign" or "strange." As markers of long-term change, Taylor claims that place names are symbols signifying such socio-cultural shifts as "emigrations, immigrations, the commingling of races by war and conquest, or by the peaceful processes of commerce." He also posits place names as indicators of geologic change, stating that: "Vast geological operations are still in progress on this globe; continents are slowly subsiding at the rate of a few inches in a century; while new lands are uprising out of the waters... [b]ut these changes, vast as is their aggregate amount, are so gradual that generations pass away without having made note of any sensible mutations. Local names, however, form an enduring chronicle, and often enable us to detect the progress of these physical changes." He lists a number of examples: former ports, like *Stourmouth*, *Chiselet*, and *Fordwick*, on what are long-receded estuaries of the Thames; the towns *Holbeach* and *Wisbeach*, now miles from the oceans that once gave them their names. As well, he explains, the tides no longer reach the town of *Tydd* and the "*Black Forest* in Argyle is now almost entirely destitute of trees." For Taylor, though the forces of culture and geology are ever at work altering histories and the landscape, the etymology of place names, lingering on maps and in regional parlance, illustrate what once was, even when *place* itself has been altered or erased beyond recognition.

---

4  Isaac Taylor, *Words and Places: Or Etymological Illustrations of History, Ethnology and Geography* (London: Macmillan and Co., 1865).

Landscapes of Erasure: Lost Villages of Canada, Lock 21 at Dickinson's Landing under the St. Lawrence River, Long Sault, ON, May 2012.

## THE DESTRUCTION OF PLACE AS PRODUCTION OF SPACE

The power to name *place* dwells in the domain of the socio-cultural, while the power to produce and eradicate the terrain exists in the realm of the material. Acts of geology, weather, and climate create and destroy geography in mere instants and over epochs. When a volcano along a volatile fault line deep under the Pacific Ocean erupts into an archipelago, explosions of molten rock form islands that, after millions of years, tower above the sea floor. However, as has been debated for nearly half a century by geographers and cultural theorists, the production of *place* lies elsewhere.[5] Crowned by its eventual flora and fauna, the *place* we now call "Hawaii" does not exist until the geologic collides with the socio-cultural. Conversely, if predictions about global climate change are correct and sea levels are rising, a similar set of oceanic islands might someday disappear beneath the waterline, effectively erasing *place* from the landscape.

The human potential to produce and eradicate both place and terrain now rivals the geologic. Engineering, warfare, and industries of extraction produce and remove landscape and land masses at increasingly greater scales. Just as shifts in plate tectonics over millions of years might result in an archipelago, a developer in Dubai has piled millions of tons of fill into the ocean to produce a set of islands in the shape of a palm tree, marked by the place name "Palm Jumeirah."[6] Or consider the destructive power unleashed during World War II, where atomic power destroyed Hiroshima and Nagasaki but even conventional firebombing reduced the cities of Tokyo and Dresden to ashes.[7] Despite the increasing scale of potential for material erasure of landscape and terrain, erasure of place is not a new phenomenon, as socio-cultural processes have been in constant flux from whatever point we choose to measure the start of "human history." Over time in urban, rural, and even "wild' space, rivers are diverted, ponds drained, malls constructed, casinos imploded, forests burned, crops grown, oceans

---

5  See Henri Lefebvre, *The Production of Space*, trans. Donald Nicholson-Smith (Oxford, UK: Blackwell, 1991); J. Nicholas Entrikin, *The Betweenness of Place: Towards a Geography of Modernity* (Baltimore: Johns Hopkins University Press, 1991); and Andrew Merrifield, "Place and Space: A Lefebvrian Reconciliation," *Transactions of the Institute of British Geographers* 4 (1993): 516-531.

6  Brett Hansen, "Artificial Islands Reshape Dubai Coast," *Civil Engineering* 75.8 (2005): 12-13.

7  See Jonathan Rauch, "Firebombs Over Tokyo," *Atlantic Monthly* 290.1 (2002): 22, and William W. Ralph, "Improvised Destruction: Arnold, Lemay, and the Firebombing of Japan," *War In History* 13.4 (2006): 495-522.

polluted, reservoirs created, clouds seeded, cathedrals erected, villages sacked, neighborhoods gentrified, libraries filled, satellites launched, histories forgotten, immigrants deported, businesses turned over, invasive species introduced, indigenous people displaced, and landmarks renamed.

Thus, what lexicon do we use to discuss such changes to the landscape? For example, how do we speak of the phenomenological vestiges of Kaixian, an 1,800-year old city of thousands, erased from the landscape under the waters of the Three Gorges Dam?[8] How do we classify a toponym like Kayford Mountain, in West Virginia, memorialized in place name on maps, in databases, and in collective memory, but whose referent has been entirely removed from the viewscape by a mining process called mountaintop removal?[9] What are the Lost Villages of Canada, once a series of towns, like Milles Roche, Wales, Moulinette, or Maple Grove, in Southern Ontario, if their remnants now rest seventy-five feet below the surface of the St. Lawrence River—yet their place names mark a series of islands produced by the dammed waters of the St. Lawrence Seaway?[10] Where am I, precisely, when I stand atop the stone foundation of a hundred-and-fifty year old building in St. Thomas, Nevada, a site that, until a decade ago, was submerged under thirty-five feet of water and, eighty years ago, was razed to make way for a man-made lake in the Southwestern American desert? And as if the erasure of place were not complex enough, the ghost town of St. Thomas has now returned to its place due to a decade of drought and the evaporation of Lake Mead.[11] These are the *landscapes of erasure* to which I am drawn.

## PLACE AS CULTURAL LANDSCAPE

Central to the appreciation of landscape in contemporary Western thought are the writings of John Brinckerhoff Jackson, whose contributions were foundational to what has come to be called "cultural landscape studies."[12] In the first chapter of his 1984 text *Discovering the Vernacular Landscape*, Jackson traces usage of the term to art history, which he states defined *landscape* as the "portion of land which the eye can comprehend at a glance."[13] Jackson further notes that the term did not refer to material space set out before the viewer but to the representation of space as captured by the artist. After stating that *landscape* is "a space on the surface of the earth," "a concrete, three-dimensional shared reality," Jackson calls for a new definition: "a composition of man-made or man-modified spaces to serve as infrastructure or background for our collective existence."

Much work in the field of cultural landscape studies argues for the value of the ordinary.[14] As Paul Groth argues in the first chapter of the 1997 compendium *Understanding Ordinary Landscapes*, "the high-style cathedral or office tower, as well as the Depression-era

---

8  Shai Oster, "Ancient Kaixian Last to Succumb to Three Gorges," *The Wall Street Journal Asia*, 19 November 2009: 12.

9  Ken Ward, Jr., "Using Documents To Report On Mountaintop Mining," *Nieman Reports* 58.2 (2004): 12-14.

10  See Barbara Righton, "Ontario's Lost Villages," *Maclean's* 113.18 (2000): 19, and Jeff Alexander, *Pandora's Locks: The Opening of the Great Lakes-St. Lawrence Seaway* (East Lansing: Michigan State University Press, 2009).

11  Denyse Wyskup, "Interpreting Site Formation Processes Affecting Re-emergent Cultural Sites within Reservoirs: A Case Study of St. Thomas, Nevada," MA thesis, West Virginia University, 2006.

12  See Paul W. Groth and Todd W. Bressi, eds., *Understanding Ordinary Landscapes* (New Haven: Yale University Press, 1997) and Denis E. Cosgrove, "Introduction to 'Social Formation and Symbolic Language'," in Rachael DeLue and James Elkins, eds., *Landscape Theory* (New York: Routledge, 2008).

13  See chapter one, "The Word Itself," in John Brinckerhoff Jackson, *Discovering the Vernacular Landscape* (New Haven: Yale University Press, 1984).

14  See Jackson, *Discovering the Vernacular Landscape*; Groth and Bressi, *Understanding Ordinary Landscapes*; and Dolores Hayden, *The Power of Place: Urban Landscapes as Public History* (Cambridge: MIT Press, 1997).

Hooverville hut, a farmer's barbed-wire fence, or a kitchen garden" are all components of the landscape.[15] Landscapes of erasure, as I am defining the term, are not unequivocally "ordinary." Producing them requires colossal feats of engineering: the dismantling of whole towns, the damming of major rivers, or the removal of entire mountain ranges. But the remnants, the landscapes that remain when physical production has ended, are certainly banal. In the case of St. Thomas, a row of blackened tree stumps and bleached-white foundations rest amidst an otherwise red and orange desert. Milles Roche and the other Lost Villages of Canada are invisible under the waters of a large bay, unless the viewer is prepared to dive deep. From most vantages, Kayford Mountain is merely a missing peak from the viewscape—unless you view it from above. Even then, the spread of rubble and sod often cap mountaintop removal sites to create rolling green lumps suitable for golf courses and strip malls where sublime peaks once stood.[16]

Landscapes of Erasure: Strip Mine beneath Paint Mountain Ridge, Willis Branch, Fayette Co., WV, May 2012.

## LANDSCAPES OF ERASURE: ILLEGIBLE PLACE?

How, then, do we speak of places like Kayford Mountain when the mountain is gone? The visible landscape, treated like a geologic palimpsest with a possible supermarket, health center, or gas station scribbled into the space where once a sublime peak towered above the horizon, will disclose nothing of the complex socio-cultural history embedded here. The place name, however, bequeathed centuries ago to a now-missing mountain, may linger on maps and in the toponyms marking new places produced to fill this space. Imagine Kayford Boulevard, Mountain View Health Center, Green Valley Dry Cleaners, or Peak Gas & Go, seemingly incongruous within a rolling landscape of subdivisions and lawns, but, as Taylor would suggest, markers that once there was a mountain here. ■

---

15  See chapter one, "Frameworks for Cultural Landscape Study," in Groth and Bressi, *Understanding Ordinary Landscapes*.

16  See Ward, Jr., "Using Documents to Report on Mountaintop Mining"; Erik Reece, "Death Of A Mountain," *Harper's Magazine* 310.1859 (2005): 41-60; and Erik Reece, "Moving Mountains: The Battle for Justice Comes to the Coal Fields of Appalachia," *Orion* 54 (2006): http://www.orionmagazine.org/index.php/articles/article/166/.

Anthony Easton & Rachel E. McRae

# 16. BEING WITH(IN) THE GEOLITHIC: INTERNET AS SHAMANIC TOOL

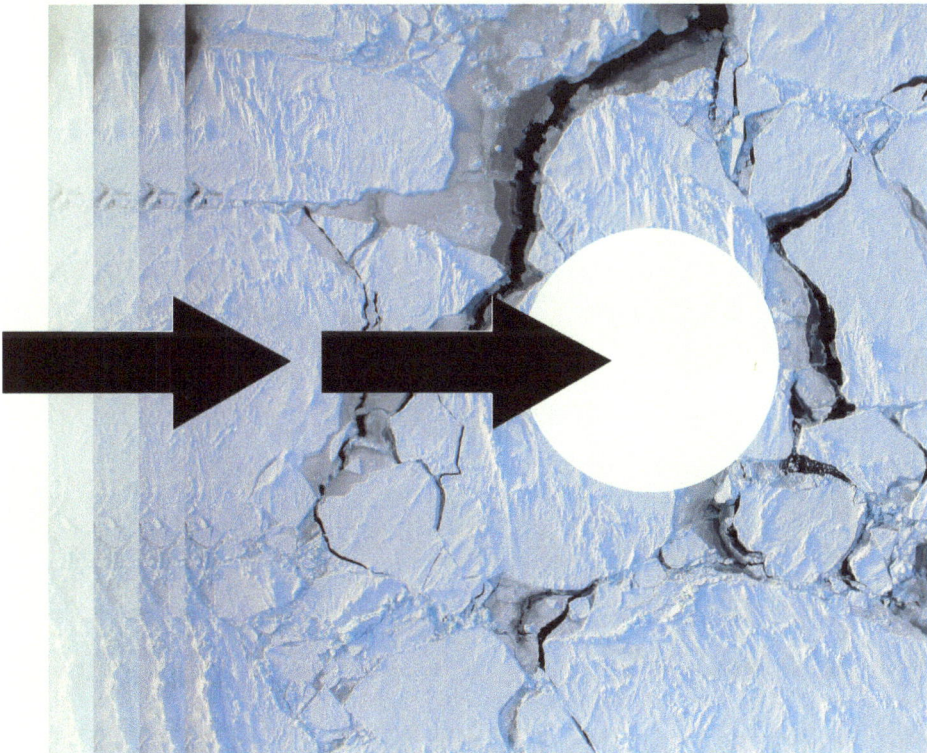

*Untitled*, Rachel E McRae, digital collage

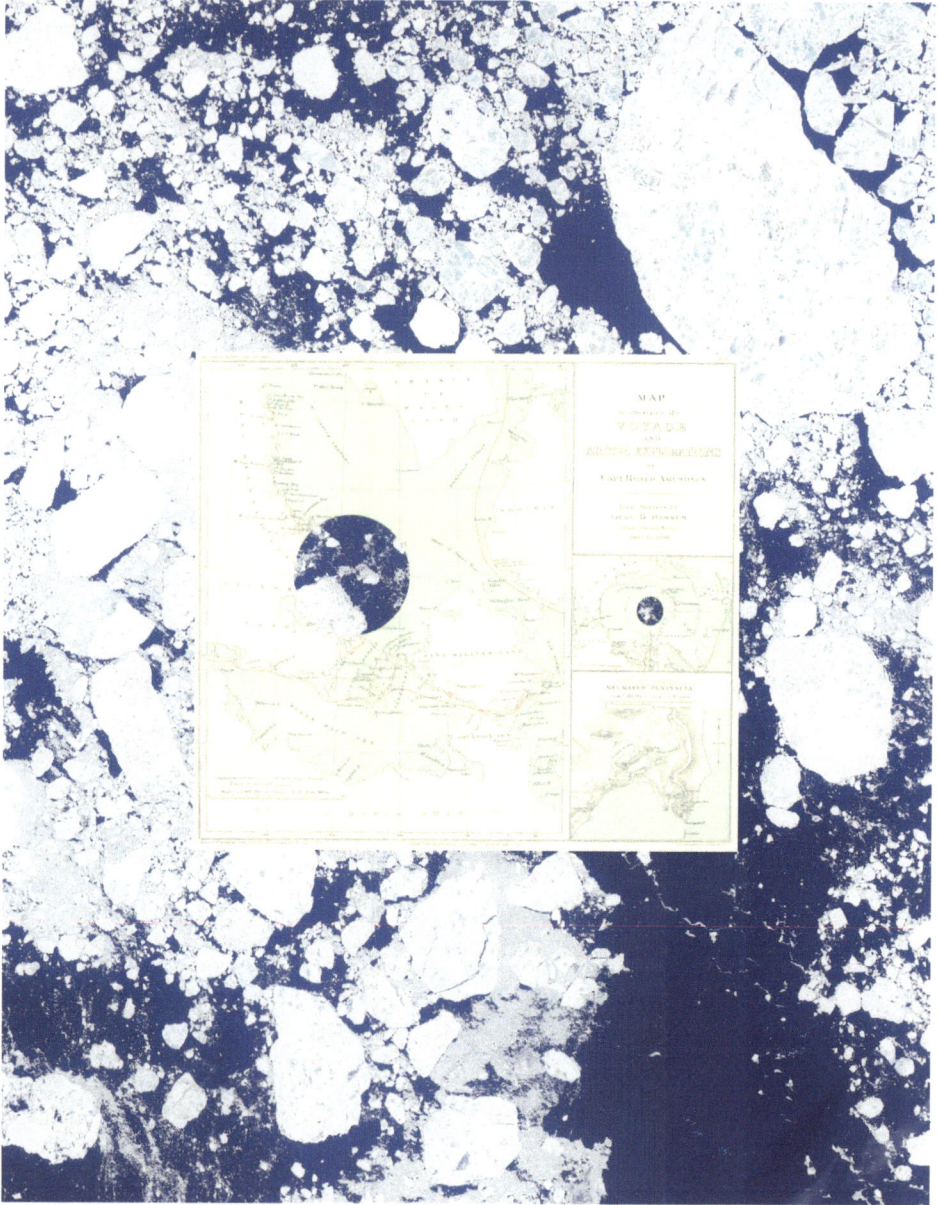

*Untitled*, Rachel E McRae, digital collage

As the medium shaping much of our current world view and mapping our interior geography, the landscape of the internet also frames our contemporary sensings of the geologic. But between the thinking and the experiencing of internet time and geologic/deep time, a fissure opens up. How do we reconcile the co-existence, both corporeal and intellectual, of two distinctly different conceptions of time and permanence: one a web-space where the longevity of a place or object is dependent upon its continuous viewership and refresh-rate, and the other a slowly shifting behemoth we physically travel across? How might an embodied knowing, a deep-knowing of deep-time, locate itself within the speed-driven space of the internet?

In 2004, the former Official Languages Commissioner of Nunavut Eva Aariak with a team including Leena Evic, the director of Pirurvik Centre for Inuit Language, Culture and Wellbeing developed a new word for World Wide Web, *Ikiaqqivik*:

> The term *Ikiaqqivik* is for Worldwide Web, and the term *Ikiaqqijjut* is for Internet. *Ikiaq* is a root word for layer. *Ikiaqqijjut* is then the instrument for going through layers, as in the Internet which is our modern instrument for layering through information near and far in an instant.
>
> *Ikiaqqijuq* was also used to describe a shaman when he traveled through layers of different worlds through his shamanic power. A quote is taken from *The Arctic Sky: Inuit Astronomy, Star Lore, and Legend* by John MacDonald: "Shamans, in their spirit flights, visit otherwise inaccessible regions, both above and below the Earth, returning with illuminating accounts of their explorations" (page 19).
>
> Our terminology development team thought the term *Ikiaqqijjut* suited how the Internet allows its users to access information without leaving their space of location. *Ikiaqqivik* refers to websites and worldwide web as it's the locations where one can browse through layers of information.[1]

With this naming, the internet is perceived as landscape in which information dwells. But it is a landscape that can be accessed from the searcher's position. The search becomes a negotiation between the seeker and content, and not one determined by technology. It is a relationship that does not place all authority upon the technology and those who supposedly own it, nor is it a linear top-down relationship that ignores the complexities of user-made content. It's a relationship with options other than being with or opting out. Hacking and excavation are legitimised. The idea of choice is introduced in a way that is not determined by the technology.

The relationship acknowledged by the Official Languages Commission's choice of word suits a format filled with user-generated content, one which is hackable. This does not ignore the lack of access experienced by many users. It does highlight the choices of navigation. Holding the methods of conveyance, shamanism is a method that consists in being able to navigate multiple temporalities. Conceiving of the internet as a shamanic tool becomes a way of being *with(in)* the geolithic.

Inuit terms for internet allow us to think about content and time in a way that is no longer singular. If we are able to see our bodies as players in this and address the content as determinate, rather than the technology, then our perception shifts radically away from the idea of technology as a bully that determines the temporality of the viewer. Understood as *Ikiaqqivik*, the internet is no longer perceived as an autonomous thing that speeds information up, but instead as a kind of shamanic conduit: non-hierarchical and continuous, which grants access to information instead of determining it. This allows us to conceive of the internet as an access-point to deep time. The world wide web becomes a conduit that allows for the coexistence of a multiplicity of temporalities, where peak and furrow are equal in value and speed is neither singular nor prioritized. The internet can be thought as something that is with the geologic, and within it.

In this way, we are able to be with the geolithic, to dig deep and sit slowly. It is no longer a matter of juggling states of being with a slow earth in a time of fast tech. In the lived experience of negotiating with the internet, these two temporalities no longer lie at either end of a chronology but as strata co-existing as multi-leveled, multivalent temporalities. Here, one is concurrently fast, slow, and all in between. The former anxiety, of how to be with something

---

1 Leena Evic, Email correspondence, 2011.

Untitled, Rachel E McRae, digital collage

slow and something fast at once, is no longer an issue. We can conceive of the coexistence of the two—not along a time-line, but as a non-linear and multi-platform web: time as shifting and linking wiki.

Linear chronology hampers and confines the deep and the old. It is a barrier that makes it difficult or at the very least terrifying to be with the geologic. Perhaps it has become easier to be cognisant of the geologic, now that the internet has removed the barrier of linear chronology. ■

Brian Davis

## 17. LAND MAKING MACHINES

Every machine, in the first place, is related to a continual material flow that it cuts into.

–Deleuze and Guattari, *Anti-Oedipus*, 1972

Cubits Gap is a major subdelta of the Mississippi River. The gap formed in 1862 after an oyster fisherman, Cubit, and his daughters excavated a small ditch in the levee between the River and the oyster-rich Bay Ronde. Intending to make an easier portage for their rowboat, they created a small crevasse. In springtime, meltwater poured through it. Six years later the crevasse was 2,427 feet wide [Ref. 1]. By 1940 a landmass larger than New Orleans had been created and the Bay Ronde had completely disappeared. Today the Cubits Gap subdelta is 40,000 acres of National Wildlife Refuge and is quickly subsiding back into the Gulf of Mexico, a microcosm of the dynamism and potentiality of the Mississippi River as the great *land making machine [LMM]*.[1]

This LMM built the Mississippi Delta over the last 5,000 years with the massive slurry loads it transports from the heart of the continent to the Gulf of Mexico. It is estimated that before 1930, this load was about 400 million tons of sediment per year [Ref. 2]. Since 1930, this load has decreased significantly. Current estimates range from 145 million to 230 million tons per year. This reduction can be explained, in part, by the 1936 Flood Control Act and the decision by Congress to reconstruct the river as a flood control and navigation system on a continental scale.[2] Despite this massive reduction, the LMM still discharges sediment at a rate roughly equal to that of the next six largest rivers in the United States combined [Ref. 3]. Most of this sediment is no longer used to continuously build the delta. Rather, thanks to the 1870's engineering of Capt. James Eads, it is shot out into deepwater in the Gulf of Mexico.[3] Since 1930 a landmass approximating the size of Delaware has been lost from the Louisiana coast [Ref. 4]. But the potential of the LMM remains.

### NEW ORLEANS: OUTPOST TO EMPIRE

Human settlement at the site of New Orleans has occurred under four different empires: Mississippian, Spanish, French, and American. During the first three, settlement took the form of commercial and military outposts rather than population centers. Each regime imposed different techniques and practices for living in the shifting, volatile geology of the region. The French and Spanish, beginning with DeSoto's ill-fated expedition and Bienville's Dilemma,[4] focused on mapping the delta and its network of lagoons, channels, islands, swamps, and forests. The result was the construction of a strategic geography of trade and defense. Though the maps belied a certain bias towards geological stasis, a wildly inappropriate assumption in

---

1  I first heard the term land making machine coined by New Orleans geographer Richard Campanella at a talk in 2010.

2  The revetments, cutoffs, spillways, and levees, as well as the modern agricultural practices being instituted at the time, greatly reduced the amount of sediment available and the hydro-electric flood control dams erected along the tributary rivers trapped much of the remaining sediment.

3  To obtain a shipping channel 28 feet deep, the depth needed for ocean-going vessels at the time, Eads proposed to build a system of long, parallel jetties at the South Pass of the Mississippi River, rather than trying to continually dredge it. His system would focus the currents of the river itself, creating enough velocity so that the river itself would scour its own channel, transporting its sediment load to the end of the jetties. To maintain this system, the jetties need only be extended periodically until deep water was reached.

4  See Richard Campanella, *Bienville's Dilemma: A Historical Geography of New Orleans* (Lafayette: Center for Louisiana Studies, 2008).

the delta, they enabled a series of settlements, battlements, and canals for transportation and drainage. In 1803, the city was important but small. Its heterogeneous population of 8,212 included French, Creoles, Spaniards, Cubans, Mexicans, Acadians, Anglo-Americans, British, Haitians, and Canary Islanders in addition to free blacks and slaves [Ref. 5].

With the advent of the American Empire in 1803, the widespread adoption of the steam-ship beginning in 1811,[5] and the water navigation boom of the 1820's-1850's, pressure built to remake the city into a population center. This was fully realized a century later with the invention of the Wood Screw Pump in 1913 and the construction of US Army Corps of Engineers flood control systems in the 1930s and 1940s. These strategies allowed the population to spread from the old levees into the low-lying swampland between the river and Lake Pontchartrain; between 1810 and 1950 the population grew from 17,242 to 570,445 [Ref. 6].

But what about that fourth empire[6]—the Mississippians? What was their regime for settlement in the Mississippi Delta? Little of their legacy is now legible: we have the fantastical Mardi Gras Indian tradition and the names of a few geographic entities, including "Mississippi." They were a fragmented and fractious society loosely associated and bound together through cultural practices, trade, and their environmental situation. The Mississippians were a "mound-building people," a fact that is particularly relevant to the discussion of a geologic turn in urban settlement. While laden with cultural significance, these mounds can also be seen as a dispersed, cellular infrastructural adaptation to the dynamic geological condition of the Mississippi Delta.

Indian Mound Park near St. Paul, Minnesota in 1898; image from the Library of Congress.

---

5 In 1811 the *New Orleans*, commissioned by Richard Fulton and leaving from Pittsburgh, was the first steamboat to traverse the entire Lower Mississippi Basin all the way to New Orleans.

6 Here, I use the term loosely; the point I illustrate is that the settlement at New Orleans was part of a unified regime of imperial scale in terms of territory and infrastructure.

The imperial capital of the Mississippians was Cahokia. At its peak, it was the largest settlement on the continent north of Tenochtitlan, present-day Mexico City. Its most visible legacy is the famous mounds located outside of present-day St. Louis. They are an exemplary instance of the mounds that can be found throughout the Mississippi basin. Popular and anthropological interpretations of the mounds have varied wildly through time. The most interesting theory posits that they functioned as a place of refuge during floods. In a 1927 issue of *Science*, the year of a great Mississippi flood, an article titled "Indian Mounds as Flood Refuges" reads:

> The thousands of terror-stricken people who have taken to Indian mounds to escape the flooding Mississippi waters are showing scientists how the Indians probably used these earthworks which they built in pre-Columbian days.
> "The buildings [on top of the mounds] were probably temples, altars and the habitats of chieftains," said [anthropologist] Dr. Kidder. "In time of flood a mound could accommodate the entire tribe, most of the members of which probably lived in the inundated area."
> Pyramidal in structure, but with a flat top to permit erection of buildings, the mounds are about 150 feet in diameter and some fifty feet high. They are largely confined to the flood area of the Mississippi. [Ref. 7]

This practice of mound-building varied but was endemic throughout the empire, from a few small hills near Kincaid, Illinois to the imperial complex of Cahokia to the impressive shell middens of the Louisiana Delta. They were not only burial sites, giant cosmological clocks, or the temple of the high priest; they were a multifunctional networked infrastructure–the construction of the territory as an articulated surface for resisting periodic inundation.[7] They were places of refuge and gathering during floods.

## DELTA FORMS AND DYNAMICS[8]

New Orleans is a city of two topographies—high and low. While subsidence, real estate values, unemployment and vacancy are all extremely high in the low city and aging infrastructures are failing, the opposite holds true for the parts of the city on the high ground near the river.[9]

---

7   This is reminiscent of the incredible constructions created by homeowners this spring to try and save their own homes as levees up and down the Mississippi were breached or blown, spillways were opened, and the Mississippi River reclaimed its floodplain for a time. Along with the tragic losses, inspiring stories abound of families and small communities collecting bobcats and sand bags and performing what we might call a "series of tactical operations" to try and save a house.

8   The recognition of the importance of geology to modern American urbanism dates at least to the 19th century, with the work of Olmsted and Vaux in New York. When the first professional degree program was formed at Harvard College in the Lawrence Scientific School, Nathaniel Shaler was the dean and William Morris Davis taught required classes. Both were geologists [Ref. 10].

9   In addition to the damaged and insufficient levees that allow catastrophic damage, many of the sewer, water, road, and drainage systems that enable day-to-day living are inadequate, costly, and aging. The word "infrastructure" is a modern term, first being used in 1875 to denote the "subordinate parts of an undertaking, substructure, foundation," especially those pertaining to military installations [Ref. 8]. This modern infrastructure was conceived according to the paradigm of the 18th century French state, cultivated at the *École Nationale des Ponts et Chaussees* [National School of Bridges and Roads] in Paris and writ large across the American continent by the United States Army Corps of Engineers. This paradigm assumes a relatively stable geology- pipes can be laid, walls constructed, rivers dredged and levees built according to predictable and stable geo-hydrological patterns. While appropriate in the bedrock of Manhattan or the stable substrates of much of the Midwest, this paradigm is proving problematic in dynamic settings such as the Mississippi delta, watersheds with high seasonal variability, or in seismically active areas such as the Isabella Lake dam in California [Ref. 11]. Given an economic and environmental climate of increasing volatility and an increased awareness of geological processes, how might a geological approach influence the development of a new infrastructural paradigm for living in the Mississippi Delta?

The case of St. Anthony neighborhood is instructive. The area sits six feet below sea level and sustained heavy flood damage during Hurricane Katrina. With an average annual subsidence rate of 5mm and a rise in sea level of 1-2mm per year [Ref. 8], the neighborhood may be nine feet below sea level by the year 2100. The neighborhood also suffers from high levels of residential vacancy with 38% of properties currently vacant [Ref. 9].

Despite these challenges, the city has two powerful allies: the fecundity of the delta, and the Land Making Machine. A strategy that activates these resources while drawing cues from the historical imperial regimes of the delta might offer a way to reconstruct the low city. Through the recovery and instigation of cultural practices, the low city might be reconfigured in the next generation not solely as a population center, but also as a vibrant and unique urban outpost within greater New Orleans.

## THE PROVISIONAL FOREST

*Provisional*: Of, belonging to, or of the nature of a temporary provision or arrangement; provided or adopted for the time being; supplying the place of something regular, permanent, or final [Ref. 10].

The fecundity of the delta has long been a primary economic source in the Mississippi Delta. Everything grows here, and grows fast. The high rates of vacancy might be worked with in this situation to institute a reforestation program for the low-city, which was once a forested bald-cypress swamp. The rate of growth for southern yellow pine timber species such as *pinus elliotii* suggests that within a generation, a mature forest could be ready for harvesting as timber pilings. The design of specific zones of forest could provide shaded, low-maintenance recreation areas for the neighborhood and also allow for the establishment of slow-growing bald cypress trees. Evapotranspiration through extensive forestation may also reduce the workload on the municipal pumping stations run by the Sewerage and Water Board of New Orleans. While creating transitional recreational and work spaces, the forest would eventually provide a harvest of timber piles for construction, leaving behind an open, maturing bald cypress forest.

## THE CITY AS ARTICULATED SURFACE

The Bonnet Carre Spillway six miles upstream of New Orleans is a critical component of the USACE Comprehensive Flood Control Plan for the Lower Mississippi basin. The spillway can discharge floodwater at a rate of 250,000 cubic feet per second, easing pressure on the levees at New Orleans by sending water around the city and into Lake Pontchartrain [Ref. 9]. An average opening event deposits nine million cubic yards of sediment in the spillway, sediment that must be dredged or hauled away in order to insure proper continued operation of the spillway. An opening event of this size is infrequent but regular, occurring about once per decade since the spillway was first operated in 1937.

The nine million cubic yards [Ref. 11] deposited in a single event would be enough to raise the elevation of the entire neighborhood of St. Anthony by 30 feet. Because both the spillway and the neighborhood border Lake Pontchartrain, barges could be used to cheaply transport some of this sediment. While it is unlikely that all of it could be recovered and transported economically, only a small percentage would be needed to raise strategic locations within St. Anthony enough to counteract subsidence and sea level rise and to withstand future storm surges. Sites that cluster social, educational, health and safety resources could be designed and elevated using poles from the provisional forest and sediment from the spillway.

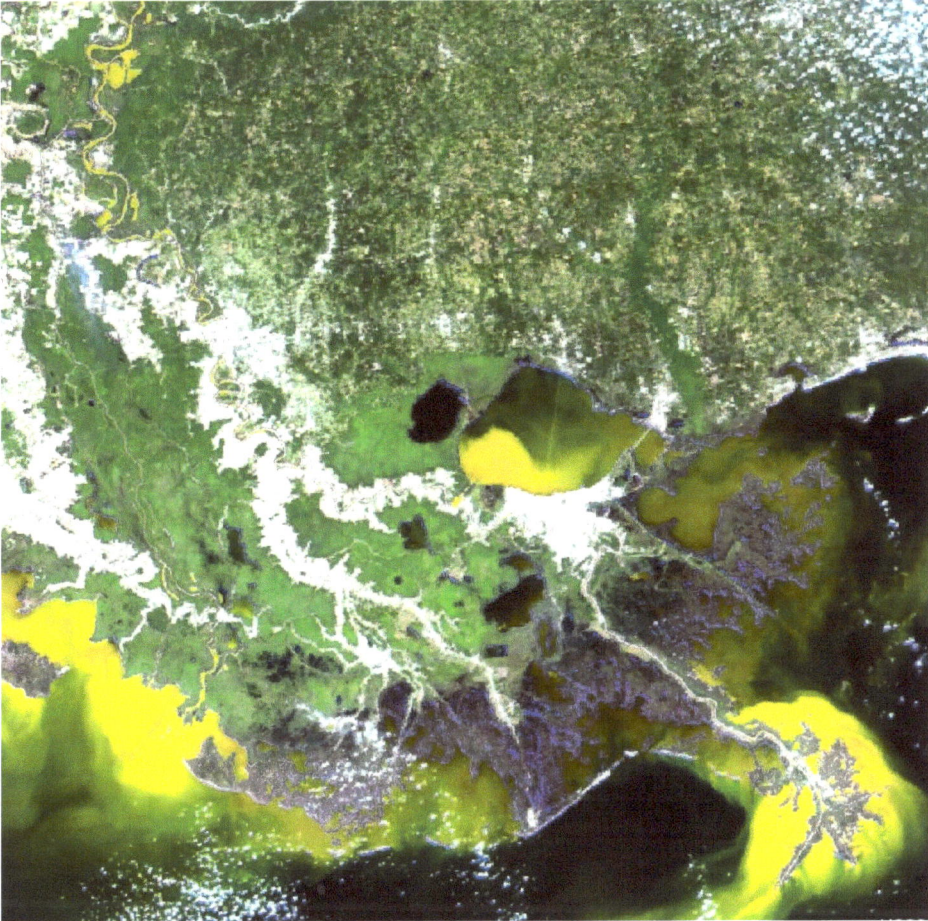

USGS satellite image of the lower Mississippi River basin during the 2011 flood; sediment plumes are shown in yellow; the intense yellow plume in the center is in Lake Pontchartrain, channeled through the Bonnet Carre Spillway.

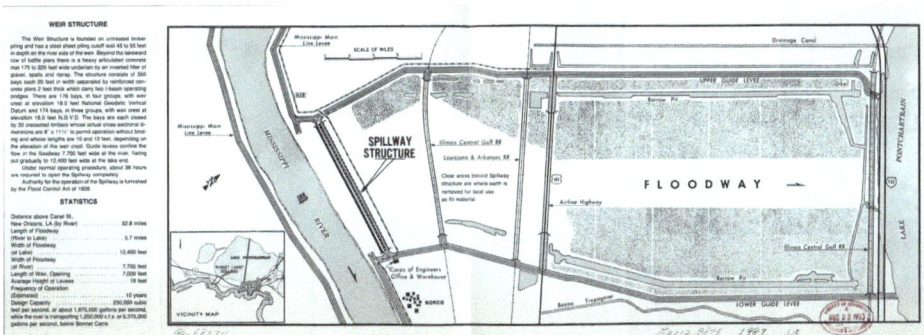

USACE schematic for the Bonnet Carre Spillway along the Mississippi River, north of New Orleans; image from the US Army Corps of Engineers.

The task of choosing which areas to raise would require involvement of local expertise and sentiment. Blocks with large amounts of vacancy, current city parks, and abandoned school buildings might be ideal candidates, owing both to an ease of acquiring the land and their prominence within the community. A network of work parks and field schools might be developed for the neighborhood, targeted to become elevated places within the city where experimentation, research, and construction is undertaken in partnership with primary schools, recreational centers, local universities, and public agencies.

In New Orleans, new urban landforms-constructed from timber pilings harvested from the provisional urban forest and sediments deposited in the Bonnet Carre Spillways would provide high ground where social, economic, and educational resources could be clustered. The cellular nature of the constructions would allow them to spread and agglomerate over time as more forest matures and the Land Making Machine deposits more sediment. Within two generations, the result would be a synthetic Bayou Urbanism—the construction of the city as an articulated surface for resisting periodic flooding. ∎

The articulated concrete surface at the base of the Bonnet Carre Spillway flood bays; image from the US Army Corps of Engineers.

References

1  Geoscience and Man, Volume XVI, The Mississippi River Delta, Legal-Geomorphologic Evaluation of Historic Shoreline Changes, David Joel Morgan. School of Geoscience, Louisiana State University.

2  "Current and Historical Sediment Loads in the Lower Mississippi River," Thorne, et al. Report by European Research Office of the US Army, Final Report. July 2008.

3  USGS Suspended-Sediment Database, Daily Values of Suspended Sediment and Ancillary Data. Website. http://co.water.usgs.gov/sediment/conc.frame.html. June 6, 2011.

4  "The Mississippi River Delta," National Wildlife Federation. Website. http://www.nwf.org/Wildlife /Wild-Places/Mississippi-River-Delta.aspx. August 31, 2011.

5  New Orleans Cabildo: Colonial Louisiana's First City Government 1769-1803. Gilbert Din and John Harkins. Louisiana State University Press, 1996: 6.

6  "Population of the 100 Largest Cities and Other Urban Places in the United States: 1790 to 1990." US Census Bureau, Campbell Gibson.  June 1998. Website. http://www.census.gov/population/www /documentation/twps0027/twps0027.html. October 17, 2012

7  "Indian Mounds of Flood Refuges," Science. Volume 65. No 1688. 6 May 1927.

8  "Sea Level Rise and Subsidence: Implications for Flooding in New Orleans, Louisiana." USGS, Science Education Resource Center at Carleton College. 2003.

9  "Neighborhood Recovery Rates," Greater New Orleans Community Data Center. 1 July 2010.

10  Oxford English Dictionary.

11  Bonnet Carré Spillway. US Army Corps of Engineers, New Orleans District.

12  Official Register of Harvard University. Cambridge Station. 24 March 1903.

13  United State Military Report, "Corps of Engineers studies risk of fault under Lake Isabella Dam," October 30, 2009.

"So let me say that right now I am experiencing simultaneously the rise, apex and decline of the so-called opulent societies, the same way a rotating drill pushes in an instant FROM ONE MILLENNIUM TO THE NEXT AS IT CUTS THROUGH THE SEDIMENTARY ROCKS OF THE PLIOCENE, THE CRETACEOUS, THE TRIASSIC."

—Italo Calvino, from "The Petrol Pump"

refinery outside Salt Lake City, UT
image smudge studio 2010

Brett Milligan

# 18. SPACE—TIME VERTIGO

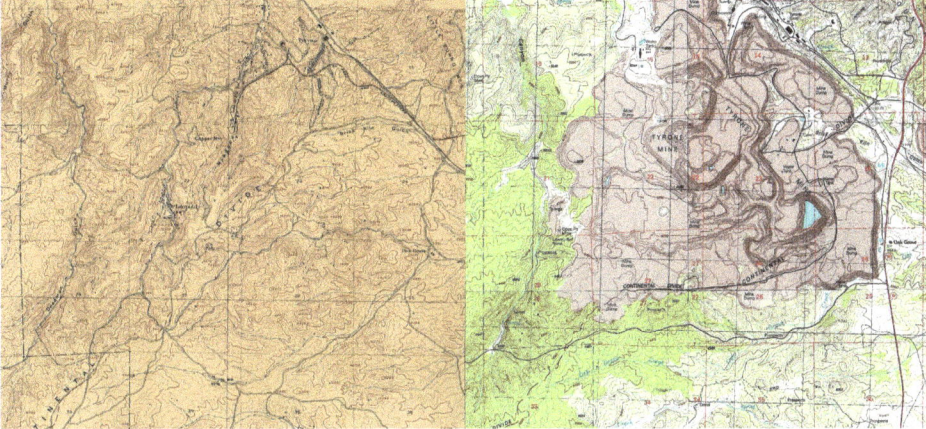

Fig 1: 1915 and 1992 USGS surveys of Tyrone, New Mexico Courtesy USGS

## SURVEYING NEW TERRAINS

In 1897 the United States Geological Survey (USGS) began the task of surveying the entire nation. From 1897 to 1992 the survey produced over 55,000 7.5-minute topography maps, creating the only uniform map covering the entire extent of the lower continental 48 states in detail.[1] These surveys became the definitive source for *remote sensing*[2] the country's lands and what was being inscribed upon them. In contrast to its strictly 'geologic' surveys, inventories of subsurface mineral deposits, USGS topographic maps gave equal consideration to surficial landscape features (landforms, rivers, and streams) as well as our cultural alteration of those forms and surfaces. Because the topographic surveys mapped the cultural terrain, they needed to be periodically updated to include how our use of the landscape was changing it.

Figure 1 shows two USGS topographic maps depicting the same location and extent of Basin and Range territory in southwestern New Mexico. The first was surveyed around 1915 and the second around 1992. In the interval between them, the survey's conventions for drawing the landscape remained largely the same. Differences in style are subtle in comparison to the massive physical changes in the terrain they depict. What in 1915 was a landscape of undulating hills, underground mine shafts and the sprawling town of Tyrone has vanished within the manufactured geography of an open-pit copper mine. In the process, over seven contiguous miles of the former Continental Divide was displaced from its former location.

---

1  The task set by the United States Geological Survey, as officially stated was, the "classification of the public lands, and examination of the geological structure, mineral resources, and products of the national domain": USGS, http://ncgmp.usgs.gov/ ncgmpgeomaps/geomaphistory/.

2  Remote Sensing is the act of obtaining information about an object or phenomena without actually making physical contact with it. More specific to geography and the study of landscapes, remote sensing is: "the process of collecting and analyzing data about the earth's environment from a distance, typically from an aircraft or satellite" James La Gro, *Site Analysis* (New York: John Wiley and Sons, 2001).

Overlays of surveys such as these provide time-lapse snapshots of the arrival of the Anthropocene, or the era of the "human-influenced environment."[3] At this distant scale (1:24,000 in the original maps), we can *see* the anthro-geology of open-pit mining occurring at a pace exponentially faster than the relatively unchanged terrain surrounding it; an anthro-geology that comes with its own repertoire of landscapes and processes.

## ENGINEERED GEOLOGIES

The amount of copper recoverable in the mining of porphyry deposits such as Tyrone's is typically less than 0.8% percent of the total quantity of rock being mined. The economy of scale required to make these operations profitable is manifest in the monumental re-sculpting of the terrain it works with. As anthropogenic facsimile of a geologic eruption, unknown numbers of nitroglycerine explosives were detonated at the Tyrone Pit over 25 years (1967-1992) to blast apart the Precambrian to Quaternary aged formations. Once fractured, the rock was shoveled into a fleet of 50' tall trucks, each capable of hauling 350 tons of material to one of two

Fig 2: Overlays of historic and contemporary surveys of the continental divide as it traverses the Tyrone Open-Pit Copper Mine (2009)
Image by author using overlays of USGS surveys

---

3  As the scientists who are currently defining the new Anthropocene state: "The scale of change taken place so far, or that is imminent or unavoidable, appears to have already taken the Earth out of the envelope of conditions and properties that mark the Holocene Epoch... the Anthropocene represents a new phase in the history of both humankind and of the Earth, when natural forces and human forces became intertwined, so that the fate of one determines the fate of the other." See J. Zalasiewicz, M. Williams, W. Steffen, P. Crutzen, "The New World of the Anthropocene," *Environmental Science & Technology* 44.7 (2010): 2228-2231. Like all other geologic eras, the Anthropocene is being defined (and contested) by scientists via evaluation of changes in the stratigraphy of the earth.

Fig 3: Space, time and empirical gaps appear in modeling terrain of the continental divide (red) as it passes through Tyrone. Stream and watershed divisions draped onto the USGS's triangulated surface display gross and informative inaccuracies. Operating as a repository of superseded surveys and dynamic geologic forms, this is a map of the Anthropocene. GIS data courtesy of USGS

accelerated geologic processes. Higher grade ores were taken to the mine's mill where they were pulverized into a sandy slurry for extraction of copper through floatation and smelting. Lower grade ores were heaped into stockpiles towering over 300 ft. tall around the pit at their angle of repose. The stockpiles were (and still are) irrigated with a percolating solution of sulfuric acid that leaches the copper deposits from the rock. This *pregnant leachate* is collected from ponds at the stockpiles' bases to enter the Solution Extraction—Electrowinning Process (SX-EW) which is able to transform liquid copper into solid, 99.9% pure sheets.[4]

At sites such as these, copper is amalgamated and then dispersed into a global urban network. As it settles into that network, it is re-deposited in subcutaneous locations, such as the internal wiring of buildings, appliances, computer hardware, and power tools. It is more than likely that there are former pieces of Tyrone's geology now embedded in either your automobile, household or surrounding community, regardless of where you live.

---

4  As described by FreePort-McMoRan: "This copper-bearing solution is captured and sent to a solution extraction/electrowinning plant (SX/EW), where the copper is concentrated in an 'electrolyte' solution, and then electroplated (or 'electrowon') onto sheets of very pure copper, ready for market". See FreePort-McMoRan, "Mining Reclamation in New Mexico," http://www.fcx.com/envir/pdf/brochure/reclamation_ NewMexico.pdf. I observed a mine employee demonstrate the efficacy of the process by placing a stainless steel wrench in one of these SX-EW baths. Like a magic trick, he held it in there for just a couple seconds before pulling it out fully coated with copper (Phelps Dodge's Morenci Mine Tour, 2006).

## INCIDENTALLY SHIFTING THE CONTINENTAL DIVIDE

Within the expanse of the Tyrone mine, the exact location of the Continental Divide has become rather tenuous and mutable. If we trace where the former divide was surveyed to have been in 1915 and superimpose it onto a current aerial, it shows a hydrological line defying all rules of topographic sensibility as it diagonally waltzes up the sides of geometric mesas and whimsically meanders into the depths of an open-pit (Figure 2). If Tyrone's manufactured topography is modeled in three dimensions with the former line of the continental divide draped upon it (Figure 3), its path reads like the soft, malleable indexes of space-time seen in Salvador Dali's *The Persistence of Memory* (1931). Only the conceptual trace of the former Great Divide, the division between Pacific and Atlantic watersheds as seen and recorded a century ago, remains. The cultured geology that rapidly superseded what was there is persistent and irrevocable. If one were to try to push all the earth back into the open-pit at the same ambitious rate at which it was pulled out (a span of only 25 years), it would take an additional 125 years (and $1.5 billion) to replace the 4 to 5 billion tons of material removed from the hole.[5]

Tyrone's geo-displacements provide a distinctive 'landmark' along the Continental Divide Trail; a marking of the land where human agency is predominantly expressed in the altered stratigraphy of the Earth's crust (Figure 4). Yet the new path of the Continental Divide was not intentionally shaped. Rather its epic reshuffling is a mere by-product of industrial

Fig 4: Tyrone open-pit copper mine, photographed in 2001. Many of the mesa-like forms surrounding the pit have since been reshaped and reclaimed since the making of this image. Image courtesy of aerial photographer Jim Wark

5  "Waiver Modification Application, Freeport McMoRan Tyrone Inc. Tyrone Mine Permit GR010RE: Open Pits and Interior Stockpile Slopes," July 2010: http://www.emnrd.state. nm.us/MMD/MARP/permits/documents/GR010RE_20100728_Tyrone_Waiver_Request_Rev_10-1.pdf.

Fig 5: The acid lake within the Tyrone Pit, maintained at a minimal depth via extensive pumping and detoxifying infrastructure. Image courtesy of aerial photographer Jim Wark

engineering. But more oddly, this shift of the Divide is essentially ignored in current field guides for the Continental Divide Trail. This segment of it is only mentioned as a required highway detour, which seems a missed opportunity to observe a seemingly incongruous and cognitively dissonant encounter along the "wild and remote" landscapes of this "backcountry trail." For at this location, one encounters an entirely new watershed detonated into the topographic seam of the continent.

Before the Tyrone mine was here, precipitation falling on this landscape was shunted away by gravity either towards the Pacific or the Atlantic Ocean, depending upon which side of the Divide it fell on. The novel tectonics of mine operations inverted that hydrologic pattern, whereby water falling on or deliberately applied to the 5,000 acres of the open-pits and surrounding stockpiles is contained and recycled within a choreographed system of ponds, pipes and pumps. Additionally, the main pit is sufficiently deep to form a "hydraulic sink" that pulls all regional groundwater towards itself.[6] If this water was not routinely pumped out of the pit (which increases the sink effect) it would fill with a lake containing 2200 mg/l sulfate with a

6  Freeport McMoRan describes Tyrone's hydraulic sink as follows: "As a result of mining, the bottom of the pit has advanced below the water table. Concurrently, to maintain dry working conditions, groundwater that flows into the pit is collected in sumps and is removed and used in Tyrone's operations. The collection and pumping of groundwater that enters the pit lowers the water table surrounding the pit creating a localized cone of depression or drawdown within the groundwater system. This response is similar to that of a well. Under these circumstances, the term given to the pit is a 'hydraulic sink' since the pit bottom represents the lowest groundwater elevation. Measurements of the groundwater elevation in piezometers and wells indicate the configuration of the water table around the pit (hydraulic sink). Within the hydraulic sink, groundwater flows into the pit. This area is also referred to as the pit capture zone. Tyrone's operational discharge permits and the closure permit (DP 1341) require Tyrone to continue to pump the pit water and maintain the hydraulic sink": "Waiver Modification Application, Freeport McMoRan Tyrone Inc., Tyrone Mine Permit GR010RE: Open Pits and Interior Stockpile Slopes", July 2010: http://www.emnrd.state.nm.us/MMD/MARP/permits/documents/ GR010RE_20100728_Tyrone_Waiver_Request_Rev_10-1.pdf.

pH of approximately 3.5[7] (Figure 5). We know this from inadvertent "pit lake" experiments at the Berkeley open-pit mine in Butte, Montana. When the Berkeley mine closed in 1982, its dewatering pumps were also turned off. Since then the pit has continued to fill, forming the largest Superfund body of water in the United States.[8] These waters, corrosive and fatal to most life forms we know, are giving rise to entirely new microorganisms that are quickly adapting to life in the sumps of the Anthropocene.[9]

Parts of the Tyrone mine are currently being reclaimed. Tailing impoundments and stockpiles have been recontoured and covered with soil and vegetation for their approved post-mining land use as 'wildlife habitat;' a herculean conversion of disparate land forms and uses performed so quickly that it makes the terrain seem like geologic stage sets.[10] However, the open-pit will remain as a monument of unreclaimability, and Freeport McMoRan will continue to pump, treat and contain water within the mine for at least another 100 years, if not in perpetuity to prevent the formation of spontaneous toxic lakes and streams.[11] The residual geology of mining has led to the design and management of a watershed with as much permanency as the path of the Continental Divide.

The Santa Rita pit as observed in 1961 (left) and 2006.
1961 image courtesy of New Mexico State University Library Special Collections

## ACCELERATION AND DISPLACEMENT

Aside from the remote sensing that maps and aerial photographs can provide, we are still left with what it is actually like to physically be with such anthropogenically accelerated geologies. If we were to travel just 20 miles northeast of Tyrone, we would encounter the Santa Rita Copper Mine. Without surrounding landmarks to tell us otherwise, we might actually feel that we were still at Tyrone, as the Santa Rita Mine is remarkably similar in its massive, constructed

---

7  Reported in "Water Treatment as Mitigation in Pit Lakes," *Southwest Hydrology*, September/October 2002: http://www.swhydro.arizona.edu/archive/V1_N3/feature4.pdf.

8  U.S. Environmental Protection Agency, "Superfund Program: Sliver Bow Creek/Butte Area," July 2012: http://www.epa.gov/region8/superfund/mt/sbcbutte/. An illustrated diagram of the hydrological sink at the Berkeley pit can be viewed at http://www.pitwatch.org/water.html.

9  For more information on these *extremophiles* see "Researchers Hope Creatures From Black Lagoon Can Help Fight Cancer," *Wired Magazine*, 27 August 2007: http://www.wired.com/science/planetearth /magazine/15-09/ff_lagoon.

10  See "What Reclamation Means," in FreePort-McMoRan, "Mining Reclamation in New Mexico," http:// www.fcx.com/envir/pdf/brochure/reclamation_NewMexico.pdf.

11  FreePort-McMoRan, "What Reclamation Means" (see above): "mineral-bearing mine water will be captured and treated for a period of 100 years (or longer), once mining has ended. In the meantime, impacted waters are captured and used as make-up water in current operations" (3).

form. Santa Rita also had a company town (Santa Rita) that was erased by the same land use for which it was founded. Where that town once was, there is now a vacuous and dusty air space above a mile-and-half wide pit. One can stare out at that space from the rim of the pit at a marked overlook on highway 152.

Harrison Schmitt, a geologist and Apollo 17 astronaut who walked on the moon, was born in the former Santa Rita Hospital. After completing his outer space mission, Schmitt joined with other Santa Ritans to re-inhabit the void of their birthplace by forming the *Society of People Born in Space (SPBS)* in the mid-1970s. The group was dedicated to celebrating their shared birth place identity and would get together annually to explore the gaps between the terrain they used to inhabit, and the non-entity it is now.[12] Given that the town of Santa Rita was fully dismantled by the early 1960's, the number of those who retain physical, embodied memories of the town grows ever smaller.

Local historian Terry Humble is a member of the SPBS and worked for the Santa Rita Mine for over 30 years. One room of his home is entirely dedicated to historical documentation of both the mine and town of Santa Rita. While sitting in this room full of all sorts of material artifacts, Terry informs me that his house (currently in the town of Bayard) is a Santa Rita original that was relocated here more than fifty years ago.[13] He mentions this while I'm staring into an 8-foot black-and-white panoramic photograph of the still-intact town of Santa Rita, looking at a blackish dot that could have been his house. In the background of the photograph a spray of dust can be seen escaping from the expanding pit, obscuring the landscape behind it; dust that is difficult to locate amongst active fault lines between mind, earth and flesh. As Robert Smithson observed, it is at such indeterminate points where those fault lines intersect that friction worth noticing seems to arise.[14] The novel forms and displacements of the anthropocene occur not only in earthen piles and pits, but in the granular mental strata trying to find the angle of repose within its new terrains. ∎

---

12  Dept. of Sociology, Humboldt State University, Community Studies Series, Report No. 4: "Santa Rita, New Mexico Community Report," 2003, http://users.humboldt.edu/slsteinberg/documents/SantaRita.pdf.
13  Conversation with Terry Humboldt, 2005.
14  "The refuse between mind and matter is a mine of information…Mind and matter get endlessly confounded": Robert Smithson, "A Sedimentation of the Mind: Earth Projects," in *Robert Smithson: The Collected Writings* (Berkeley: University of California Press, 1968), 107 [100-113].

Chris Taylor

## 19. FERTILIZING EARTHWORKS[1]

*Remains of Oficina Salitrera Prosperidad, Chile, 12 October 2007.*
*Photo credit: © 2012 Chris Taylor*

Earthworks map the intersection of human construction and geomorphology. They begin with land and extend through the complex social and ecological processes that create landscape. Propelled by the legacy of John Brinkerhoff Jackson, our understanding of earthworks as sites where the human and the geologic intersect has expanded to encompass human settlements, monumental artworks and industrial installations, as well as traces of their construction and decay, ranging from geologic material and weather to cigarette butts and hydroelectric dams.

Since humans took an active role in cultivating food and medicine, nearly 10,000 years ago, we have been involved in shaping the surface of the earth and learned quickly the importance of maintaining soil fertility by adding available organic nutrients to the ground. Human development occurred over millennia precisely because of our ability to sustain ourselves in the places where we lived. Understanding the chemical make up of soil fertility began with the identification of the key ingredients: phosphorus in 1669, nitrogen in 1772, and potassium in 1807. Nitrogen, while comprising 78% of the earth's atmosphere, was the most difficult to isolate. Feeding the biological hunger for usable nitrogen to create chemical fertilizers required a source for naturally occurring nirates—nitrogen-oxygen chemical units accessible to reaction. The arrival of industrialization, ushering in the Anthropocene, is marked by the human ability to move vast quantities of geologic material.

Terrestrial nitrates were historically found in accumulations of guano and mineral deposits within the Atacama Desert, the primary global source from the mid 1800s to World War I. The value of Atacama nitrates motivated the War of the Pacific (1879-1883), which resulted in the redistribution of territory from Peru and Bolivia to Chile. Sodium-nitrate, or Chilean saltpeter, is an essential ingredient of both fertilizer and explosives. The global redistribution of nitrates from the Atacama in the late nineteenth century marks a turning point in the ecological balance of the planet. A movement from local ecological cycles where the fertility of a region is measured by the ability to retain and reinvest nutrients into the soil, to industrial ecological cycles that measure yield productivity with the quantity of imported nutrients. The rise of industrial agriculture and world war emanate from nitrate mining in the Atacama. The creation of the Haber-Bosch process in the 1910s provided a source of synthetic "fixed" nitrogen causing the commercial viability of Chilean saltpeter to evaporate and its extraction to subside. Shortly thereafter, the mining installations were largely abandoned, scrapped, or repurposed. What remain marks the geologic and temporal displacements of our time. ∎

*Remains of Oficina Salitrera Prosperidad, Chile, 12 October 2007.*
Photo credit: © 2012 Chris Taylor

*Remains of Oficina Salitrera Prosperidad, Chile, 12 October 2007.*
Photo credit: © 2012 Chris Taylor

Remains of Oficina Salitrera Prosperidad, Chile, 12 October 2007.
Photo credit: © 2012 Chris Taylor

Remains of Oficina Salitrera Prosperidad, Chile, 12 October 2007.
Photo credit: © 2012 Chris Taylor

*Remains of Oficina Salitrera Prosperidad, Chile, 12 October 2007.*
Photo credit: © 2012 Chris Taylor

Remains of Oficina Salitrera Prosperidad, Chile, 12 October 2007.
Photo credit: © 2012 Chris Taylor

Remains of Oficina Salitrera Prosperidad, Chile, 12 October 2007.
Photo credit: © 2012 Chris Taylor

1 Images for this photographic essay were created during *Incubo Atacama Lab* (http://earthworkslab.org/
atacama_lab/") in 2007 when Chris Taylor brought the interpretive frame and working methods of *Land
Arts of the American West to Chile* (http://landarts.org).

# SECTION 3: FROM PERIPHERY TO CENTER: ARTISTS MAKE THE GEOLOGIC NOW

Susannah Sayler + Edward Morris / The Canary Project

## 21. BLANK STARE
### (10 IMAGES AND CORRESPONDING NOTES ON THE PENSIVE IN PHOTOGRAPHY AND ITS UTILITY IN THE FACE OF CATASTROPHE, AS EXCERPTED FROM *A HISTORY OF THE FUTURE*)

**1.**

Most of us barely know how to read the land. But once you start to learn, even a little, you are confronted everywhere with absence. You are drawn back into the past, and this seems, at first, the only direction you can go. You see glaciers in the striations on the stone, the flatness of the valley at the foot of the mountain, the melt water, the rocks left in a field thousands of years ago by the withdrawal of a nearly unimaginable force. Yet, feeling the past in this way is like taking a few, hard steps up a hill, gaining elevation from which you ultimately hope to see in all directions, even the future...

**2.**

Each photograph in this essay is a landscape. Each was taken between November 2005 and June 2010 under the aegis of The Canary Project, a collective that produces art and visual media that deepen public understanding of climate change. In keeping with this utilitarian promise, the photographs have been exhibited on the sides of busses, on billboards, in school presentations, in science museums, city halls, as well as in art museums and galleries.

The malleability of the photograph, its adaptability to various vernaculars and its general legibility, is one of the main reasons we decided to work in the medium given our intent to impact a broad public. Everyone feels comfortable approaching a photograph and has an idea what to do with it, but not everybody knows what to do with a painting or a sculpture or a video.

The meaning of our images depends on their context within a specific mode of exhibition and the position of that mode within a larger discourse about climate change. This discourse has many registers: scientific, journalistic, activist, and artistic. In the aggregate, the photographs form an archive and can be positioned as evidence (i.e. they can serve one register of the discourse or another, depending on how and where they are shown).

And yet, in their idealized state, stripped of context and viewed individually, the same photographs form a blank stare. Liberated in this way, they present a much bigger challenge to the viewer as their meaning is more indeterminate. In this state the photographs float and are pensive. They are shameless in their repose, even in the face of catastrophe.

**3.**

Pensiveness is not some special characteristic of our photographs. Pensiveness is at the core of all photographs, which despite the digital age, continue to carry the aura of a reminiscence as their ineluctable modality. They are an index. They point to the past, or more precisely to a discrete moment of the past. This, then, there. Photographs will not lose this connotation so long as they are the primary method for recording and sharing everyday events.

1. Glacial, Icecap and Permafrost Melting XLVII: Cordillera Blanca, Peru, 2008
Archival Pigment Print, 40"x50"

2. Extreme Weather Events III: Plaquemines Parish, Louisiana, 2005
Archival Pigment Print, 40"x50"

In the sequence that we put together for this book, we focused the innate pensiveness of the images on a topic that we believe is absolutely central to understanding climate change, but which may appear tangential: appreciating the vast differential between human history and geologic history.

**4.**

> "It makes you schizophrenic. The two time-scales—the one human and emotional; the other geologic—are so disparate. But a sense of geologic time is the most important thing to get across to the non-geologist . . . . A million years is a small number on the geologic scale, while human experience is totally fleeting—all human experience, from its beginning, not just one lifetime. Only occasionally do the two time scales coincide."
>
> (Dr. Eldridge Moores, geologist, quoted by John McPhee in *Annals of the Former World*, p. 458)

These two histories, the geologic and the human, coincide decisively in climate change. It has the potential to be the final chapter in human history, but geologic history has no comprehensible end.

**5.**

The photographs are from Peru, Antarctica, Niger, New Orleans, and The Netherlands. The very names of the places connote a particular history, a part in the human story. Obviously, we cannot capture that history or explore it any detail in such an essay, but we can encourage the imagination of it. Because photographs have reminiscence as their ineluctable modality, they stimulate the imagination (What there? What then?).

But more than this: because photographs trigger a memory reflex, the scene they depict inevitably takes on a personal significance that is like a shadow and cannot be fixed precisely. This is an unconscious mechanism, silently nagging the viewer (*"Where does this fit in my life? Was I there?"*).

The places are also, of course, points on the earth, coordinates in what we call our environment, that which envelopes us. In this sense they are part of a different history, but one that is much harder to comprehend, the story of a planet that has existed for more than 4 billion years before humans. This too is part of our collective unconscious.

**6.**

The photograph, whatever its limitations, has one distinct advantage as a medium to encourage mediation on the deep past: Its momentary nature (the thinnest veneer of time available: the instant) sets in stark relief the time-based phenomenon we hope to elucidate. Always a memory, but in this case, a memory that extends beyond the realm of human experience.

**7.**

In the field, we had two choices: 1) dramatize the landscape in some way, such as indulging in a myth about damaged earth, in an effort to activate feelings of indignation, disgust, wonder, or dismay (whether or not this motivation is frankly admitted or not); 2) confess to a feeling of incomprehension, disorientation, loss of scale, seduction, and panic. One is a gesture of mastery, and the other of submission.

*3. Disrupted Ecosystems XXI: Monte Verde Cloud Forest, Costa Rica, 2006*
*Archival Pigment Print, 40"x50"*

*4. The Oostvaardersplassen Nature Preserve, The Netherlands, 2010*
*Archival Pigment Print, 40"x50"*

Most photography dealing with environmental themes employs some version of the former strategy. You can see this technically in the color saturation, the single-minded themes, the bold compositional choices which preference symmetry, and an elevated perspective that is dehumanizing and gives the impression of omniscience. You can also see it in the way the work is framed and discussed, usually in the vein of heroism. In other words, most landscape photography has still not escaped the long shadow of Ansel Adams.

**8.**

"Parallelism between stepping back in time (Historicism) and moving out in space (Exoticism) . . . . The sensation of exoticism: surprise. Rapidly dulled…I conceive otherwise, and, immediately, the vision is enticing . . . ."

"The sense of a non-anthropomorphic nature, a nature that is not superhuman but ex-human and from which all humanity—strangely!—is derived—this sense of nature's exoticism only emerged from the understanding of the forces and laws of nature."

(Excerpted from, Victor Segalen, trans. Yael Rachel Schlick, *Essay on Exoticism: An Aesthetics of Diversity*, pp. 48 and 22)

Adherence to the lofty formal techniques of Adams' is not incidental; it is part and parcel of a philosophical outlook that regards Nature as transcendental, "a religious idea" (to quote Adams himself). Most contemporary landscape photographs tend to be more self-consciously ambiguous in their statements of purpose, yet still operative is the underlying dogma that nature and man are opposed and that, as such, we are damaging Nature. This leads to a simplistic symbolic order, which is often evident in the compositional structure of the photographs. More significantly it also leads to adamant disavowals of any political intent or responsibility to the works. Such work creates a (visual) sanctuary and a locus of mourning that dispenses with any need to act. It assumes no future or at best longs for a future that is conservative, a future where wild places are preserved with only privileged access.

Against this tendency, we felt a perverse attraction to the following idea of nature that is even older and more out-of-vogue:

To see landscape thus, as something distant and foreign, something remote and unloving, something entirely self-contained, was necessary. . . . For we began to understand Nature only when we no longer understood it; when we felt that it was the Other, indifferent towards men, which has no wish to let us enter, then for the first time we stepped outside of Nature, alone, out of the lonely world. . . . Nature was more permanent and greater, all movement in it was broader, and all repose simpler and more solitary.[1]

Whatever this romantic perspective lacks in terms of sensitivity to humanity's position as part of an ecosystem or foreknowledge that we would come to call this era the Anthropocene, it is perspicacious and intellectually rigorous on one point: the supreme indifference of the inorganic world. The most unexpected phenomenon for us in undertaking this project was to travel to these places that scientists had identified as harbingers of catastrophe and finding them so devoid of any clear sign of danger. The land was just there, imperceptibly changing. ■

---

1 Rainer Maria Rilke, "Concerning Landscape," in Stephen Mitchell, Introduction, *The Selected Poems of Rainer Maria Rilke*, ed. and trans. Stephen Mitchell (New York: Vintage Books, 1989).

See http://canary-project.org

5. Arc, Dordrecht, The Netherlands, 2010
Archival Pigment Print, 40"x50"

6. De Gijster Reservoir, Biesbosch National Park, The Netherlands, 2010
Archival Pigment Print, 40"x50"

7. *Glacial, Icecap and Permafrost Melting XXXI, Antarctica, 2008*
*Archival Pigment Print, 40"x50"*

8. *Adaptation and Mitigation LVI: Reforestation and Land Restoration, Niger, 2007*
*Archival Pigment Print, 40"x50"*

9. Adaptation and Mitigation LIX: Water Storage, Lake Paron, Peru, 2008
Archival Pigment Print, 40"x50"

10. Glacial, Icecap and Permafrost Melting XXXVI: Bellingshausen Base, King George Island, Antarctica, 2008
Archival Pigment Print, 40"x50"

image: Jamie Kruse, Sudbury Superstack, Sudbury, Ontario, 2011

# WE ARE PART OF AN ONGOING GEO-COSMOLOGICAL CONVERGENCE

Brooke Belisle

## 21. TREVOR PAGLEN'S FRONTIER PHOTOGRAPHY

*Anasazi Cliff Dwellings, Canyon de Chelly,* 2010
From Trevor Paglen, *Artifacts*
[Detail, part one of diptych] C-print, 50 x 40 in.

The first photograph of Trevor Paglen's 2010 diptych, *Artifacts*, is familiar. Picturing ancient ruins nestled in the cliffs of Arizona's Canyon de Chelly, it echoes a 1873 photograph by Timothy O'Sullivan and a 1941 photograph by Ansel Adams. Paglen's photograph, like its antecedents, is rendered in sharp-focus, without color, dominated by dramatic, diagonal striations of the cliff face. It frames the archeological history of the ruin within the geological history of the canyon within the cultural history of photography.

Canyon de Chelly is actually the intersection of several canyons, formed by slow erosion over thirty million years ago. The angled striations seen in the photograph expose even older geological processes, layers of sediment compressing into strata then shifting, rising, and tilting. Between the 1st and 4th century, people now called the Anasazi settled in the canyon. They built the structure in this photograph around the year 1000, and, 300 years later, they had disappeared. Navajo tribes spreading across the Southwest discovered the ruins already 500 years old. In the 18th century, Spanish colonists ventured into the canyon and soon they were converting and killing Navajos who lived there.

United States expeditions arrived in the mid-1800s and fought to claim the Southwest from Arizona's Fort Defiance. In 1864, Kit Carson forced the Navajo of Canyon de Chelly to surrender, and marched them on The Long Walk, almost twenty days from home. In 1868, a treaty established the Navajo Indian Reservation, allowing them to return. Three years later, planning U.S. expansion, the War Department sponsored a Western survey and hired Timothy O'Sullivan.

Spacecraft in Perpetual Geosynchronous Orbit, 35,786 km Above Equator, 2010
From Trevor Paglen, Artifacts
[Detail, part two of diptych] C-print 50 x 40 in

O'Sullivan's 1873 photograph of Canyon de Chelly was intended as information, published in a bureaucratic report filled with measurements and maps. But the grandeur of his image, and survey photographs like it, helped romanticize the Western frontier and fuel American dreams of manifest destiny.

Ansel Adams came to Canyon de Chelly 100 years after the first U.S. expeditions and over 50 years after O'Sullivan. For Adams, the sharp-focused clarity of survey photographs articulated a new frontier for photography. In 1940, he organized a groundbreaking exhibition in San Francisco, presenting survey images as archeological finds that could ground a new origin, history, and aesthetic for photographic art. The same year, the Museum of Modern Art established a Photography Department, headed by Adams' collaborator Beamont Newhall. Newhall institutionalized photography as art, and wrote a book on O'Sullivan. When Adams re-photographed the canyon in 1941, he was articulating layers in the history of photography and in how the West has been seen.

When Adams came to the canyon it was a National Monument and a process of "reorganizing" was returning land to Navajo. Adams was on a commission from the Parks Service that World War II interrupted. War continued to affect the canyon. Weapons were tested in the desert, and in the late 1960s, supersonic aircraft shook cliffs so violently that ruins crumbled. When Paglen took his photograph, in 2010, he considered the history of the canyon and ruins, but also ongoing histories of imperialism and war that the beauty of ancient cliffs and stories of ancient ruins can obscure. His image adds another layer of visibility to a scene already stratified by continual change and rupture.

The second photograph of Paglen's diptych appears beautifully obscure. Against a dark background, faint gray streaks slant down from left to right, punctuated by brighter dots and marks that move against this pattern. The hazy, diagonal lines visually echo the shade, and mirror the direction, of the bands slanting down the cliff from right to left in the canyon photograph. The smaller, brighter marks also resonate with the other photograph, rhyming with the man-made shapes of the ruins' walls and windows, which interrupt the natural geometries of cliff ledge and canyon. The diptych's full title tells us what, exactly, is being juxtaposed: *Artifacts (Anasazi Cliff Dwellings, Canyon de Chelly; Spacecraft in Perpetual Geosynchronous Orbit, 35,786 km above Equator)*. In the second photograph, the streaks are stars, and the bright marks are man-made objects: dead satellites and abandoned shuttle parts.

Since the 1960s, hundreds of thousands of things launched into space have stayed there. Today, broken bits of space shuttles, paint chips, defunct satellites, and other fragmentary and abandoned objects orbit our planet as a belt of debris. This band will encircle Earth after humans no longer exist, forming not just a junk pile but a displaced fossil record. Using a digital telescope at an astronomical observatory, Paglen photographs this distant layer of detritus, at the outer edge of not only human space but human history. In order to visualize space debris, Paglen relies on complex algorithms that compile observational data and predict orbital patterns. He uses these algorithms to program the position and movement of a photographing telescope, enabling him to focus on the place an object should be and to track it, slowly rotating to compensate for the Earth's own rotation. The computational and visual technologies Paglen uses to photograph these objects in space are the same technologies that helped put them there. He uses technical capacities developed during the space age, which have reframed the scale of human action and impact, to recalibrate correlated scales of vision, representation, and interpretation.

Juxtaposing geological and astronomical perspectives, Paglen's diptych invokes an earlier moment when these perspectives mutually recalibrated the context of human history and meaning. During the Industrial Revolution, technological changes radically altered frameworks of space and time. Digging mines and canals brought new attention to the ground beneath our feet, and it appeared much older than we thought. In 1785, James Hutton argued the Earth's topology had developed in slow cycles over an immeasurable duration. In 1811, French naturalists demonstrated that patterns of strata could be read as a material timeline. If the depth at which fossils were found told when they were formed, this suggested the age of the Earth might be fathomed from physical evidence, counted off like tree rings. William Smith visualized those rings in a map using different colors to delineate how geological layers were stacked in vertical depth. In 1830, Charles Lyell insisted that the distant past could be read from the Earth's present form, its appearance a material expression of history. These early geologists reimagined the scale and process of history, directly inspiring Darwin, and positing what we now call "deep time."

As geological perspectives were shifting, so were astronomical perspectives. At the turn of the nineteenth century, William Herschel built telescopes that could see further beyond our solar system, and began to imagine the topography of the universe. Herschel suggested that instead of viewing the sky as a dome, the "vault of the heavens," we should regard it as a "naturalist regards a rich extent of ground or chain of mountains, containing strata variously inclined and directed" relative our gaze. Thinking of astronomical space in terms of geological formations suggested how, like strata, levels of astronomical depth express temporal distance. Herschel realized his telescopes not only penetrated further into space but deeper into the past, seeing light that had traveled for millions of years. His ideas initiated the concept of "deep space," provoking an expanded cosmological timeline. Echoing with contemporaneous geological arguments, his view unsettled notions of a fixed and perfect universe, arguing it was unimaginably old and vast, and still changing.

The way we visualize Earth and sky was reorganized with the emergence of photography in the nineteenth century. Flattening whatever it pictured into an image, photography seemed to represent everything visible as within our perceptual grasp—or, at least, gave new shape to that ambition. Geological surveys sent photographers to chart every contour of America's Western frontier. An international body of astronomers undertook a project to photographically map every star. Photography's visual archive promised to collect and juxtapose images of everything to reveal new patterns. As cameras were sent up in hot air balloons, then airplanes, then space shuttles and satellites, the scope of our visual knowledge seemed rescaled. Aerial photography was used by US Geological and Geographical Surveys to topologically map the country, and by astronauts to map the moon. Today, scientists study satellite images that picture the Earth as a globe and space as a set of star coordinates. As geological and astronomical points of view are increasingly displaced beyond human perspective, and aligned with the apparent objectivity of technological imaging, this threatens the imaginative recalibrations new perspectives require. Calibrating the scale at which meaning becomes visible, the range of space and time that seems to manifest itself within our look, always involves visualizing how whatever is seen extends beyond our view, to places and times we can only imagine.

Paglen presents the visual correspondence between the two images of *Artifacts* as a more profound correspondence of how broken histories, and shifting scales of time and space become visible. The shape of the canyons and the striations of its cliffs contextualize the history of the ruin; the motion of the Earth read off from streaking stars, and the scales of age and distance these faint streaks register, re-contextualize what we have left in space. These lost objects are also ruins, abandoned artifacts embedded in an ancient, astronomical landscape. The image of Canyon de Chelly invokes a history of receding frontiers, the intertwined imaging of art and science, and the repetition of colonial invasion that all find expression in the space race, and are all expressed today as power is exerted through outposts of orbiting satellites. As the abandoned ruin marks the aspirations and outer limits of one civilization after the next, dead spacecraft remind us that another culture, aspiration, and outer limit has always violently appeared, continuing even beyond the planet's own horizon.

Reactivating a mutual calibration of geological and astronomical perspective, Paglen visualizes debris circumscribing our planet as a strata of sediment where our traces collect and will outlast us. This exposes the expanding spatial and temporal removes at which human history takes place and leaves its mark. Paglen's juxtaposition challenges the idea of progress, asking how expanding frontiers might not expand our agency as much as re-contextualize our finitude and expand the gaps of time and space across which meaningful correspondences echo. Set against geological and astronomical scales that far exceed the human, artifacts of two civilizations a thousand years apart compress as traces of the same human history, and suggest the overarching ruin that all of human history must someday be. Challenging us to reframe our perspective within scales that vastly exceed us, his *Artifacts* diptych asks what traces we would leave. ■

References

Ansel Adams. 1941. *White House Ruin, Morning, Canyon de Chelly National Monument, Arizona*. Gelatin silver print, 19⅝x 14¾in.

Cuvier, Georges and Alexandre Borgniart. [1811] 1835. *Essai sur la géographie minéralogique des environs de Paris*. Paris: E. d'Ocagne.

Goetzman, William H. 2000. *Exploration and Empire: The Explorer and the Scientist in the Winning of the American West*. Austin: Texas State Historical Association.

Hutton, James. 1784. *Theory of the Earth; or an Investigation of the Laws observable in the Composition, Dissolution, and Restoration of Land upon the Globe. Transactions of the Royal Society of Edinburgh*, Vol. 1, Part 2: 4 [209–304].

Herschel, William. 1784. *Account of some observations tending to investigate the construction of the heavens. Philosophical Transactions of the Royal Society of London*, Vol. LXXIV For the Year 1784, Part 1: 437-451. London: Lockyer Davis and Peter Elmsly.

Kelsey, Robin E. 2003. "Viewing the Archive: Timothy O'Sullivan Photographs for the Wheeler Survey, 1871-74." *The Art Bulletin* 85.4 (December).

Lyell, Charles. 1830. *Principles of Geology*. London: John Murray.

Newhall, Beaumont. 1965. *Aerial Photography*. Rochester, NY: RAETA.

Newhall, Beaumont and Nancy Newhall. 1966. *T. H. O'Sullivan, Photographer*. New York: George Eastman House and Amon Carter Museum of Western Art.

O'Sullivan, Timothy H. 1873. *Ancient ruins in the Cañon de Chelle, N.M. In a niche 50 feet above present cañon bed*. Albumen print, 10.83 x 8 inches.

Paglen, Trevor. 2010. *Artifacts: (Anasazi Cliff Dwellings, Canyon de Chelly; Spacecraft in Perpetual Geosynchronous Orbit, 35,786 km above Equator)*. Diptych. Archival pigment prints, 40 x 50 inches.

Turner, H.H. 1912. *The Great Star Map: Being a General Account of the International Project Known as the Astrographic Chart*. Cornell: Cornell University Library Digital Collections.

Wilson, David M.B.R. 1976. *Administrative History: Canyon de Chelly National Monument, Arizona*. United States Department of the Interior, National Parks Service.

Wincehster, Simon. 2001. *William Smith: The Map that Changed the World*. New York: HarperCollins.

smudge studio (Elizabeth Ellsworth + Jamie Kruse)

## 22. THE UNEVEN TIME OF SPACE DEBRIS: AN INTERVIEW WITH TREVOR PAGLEN

image courtesy Trevor Paglen, *Dead Military Satellite (DMSP 5D-F11) Near the Disk of the Moon, 2010*

*Trevor Paglen's work, entitled Debris, is a series of photographs of man-made debris, garbage, and flotsam in earth's orbit. It shows spent rocket bodies, inactive spacecraft, and satellite fragments caused by collisions and explosions. Much of this debris will outlast humankind's presence on the planet.*

**SMUDGE:** We see your work as signaling what we are calling a "geologic turn" in contemporary cultural sensibility and awareness. What do you think of that perception?

**PAGLEN:** This is something that I think about a lot. One thing that interests me about this idea of thinking geologically, or using the geologic as a kind of analytic framework or approach is this: What would happen if you took geographic thinking and instead of putting it on a horizontal axis, you added a vertical axis to it, a temporal axis? You would be thinking not only about unevenness of the surface of the Earth, but also about the multiple ways in which time itself is uneven. If we go back to all the nineteenth century talk of the "annihilation of space and time," we find the beginnings of a world in which humans have reshaped time in the interest of capital and warfare. Mostly, we think about this in terms of speeding up time (increasing capitalist turnover times, labor productivity, financial transactions in the case of capital, and things like GPS targeting and hypersonic cruise missiles in the case of militarism).

But in addition to the industrial annihilation of space with time that we see, the 19th century and early 20th century marks the advent of the so-called "Anthropocene Age," a moment in earth's history when humans begin moving more sediment than traditional geomorphic processes (erosion, glaciation, etc....) In the Anthropocene, things like real-estate markets become geomorphic agents, because fluctuations in housing prices, for example, determine how huge amounts of sediment gets moved across the planet. My point is that human societies are both speeding up and slowing down at the same time. One consequence of these "anthropogeomorphic" processes is that the effects of our activities are played out over longer and longer time periods: one example is climate change: we are setting earth processes in motion that are going to play out over a hundreds, of not thousands of years. Another example is nuclear waste, which intersects with evolutionary and even geologic time. We are making things that exist on geologic time scales. For me, making this geologic turn is a way to try to come to terms with the uneven temporalities of contemporary human existence.

**SMUDGE:** Much of your work seems to be an attempt to illustrate the fact that "other worlds" coexist alongside, or interwoven within, mainstream daily life. The interweaving of worlds that you photograph reminds us of the latent or invisible forces of the geologic, which are constantly shaping the world we're in, and in ways that are often hidden or invisible. What is your motivation for making invisible worlds and forces visible and sense-able?

**PAGLEN:** I often think about this notion of geology, or geomorphology, in relation to human institutions. Consider a place like Guantanamo Bay, for example. I would submit that the reason why Guantanamo Bay still is there is because it's there. The chain link fences and the brick and mortar that the buildings are made of actually have a kind of historic agency. They actually want to reproduce themselves. We're all familiar with the 19th Century idea about the "annihilation of space with time" but the obverse is also true. Space also annihilates time. Whether we're talking about nuclear waste or Guantanamo Bay, we can see how materialities produce their own futures. This is a way in which materiality and politics intersect. Materiality is not politically neutral, so I think that you can talk about Guantanamo Bay as a political phenomenon. I think that materiality can explain some things.

**SMUDGE:** One of the realizations we are left with after viewing your space debris project is that space trash will far outlast human existence on the planet. Was such a realization an intentional provocation of that piece?

**PAGLEN:** Yes, that's an intentional provocation of that piece. And there are some other newer works that are very much about this as well. I also have a piece that came out last year, that's called Time Study, which is a series of photographs of predator drones done as albumin prints. That piece is about trying to think through multiple scales of time itself, and also through the history and the machines that we use to both produce and experience time. And the space debris project is very much about that as well. That project is trying to a create a framework that asks you as a viewer to try look at this world and imagine the time scale and the time frame that you're looking at.

**SMUDGE:** While doing research for *Making the Geologic Now*, we've encountered some people who say that that it's impossible for humans to grasp deep time. But we're starting to sense that's not actually true. We think that aesthetic works might assist humans to begin to cognitively grasp geologic time and geologic forces that shape our daily lives, but that we often can't see or are just outside our frame of reference. If we create artworks to bring focus to this reality, then people can begin to perceive it.

**PAGLEN:** Absolutely, that's one of the ways that I got interested in this question in the first place. As I was tracking secret satellites, I realized that the majority of spacecraft stay up there long after they've powered down and "died." And so, that got interesting to me. Even some of the earliest spacecraft are still up in orbit. When you start getting into the orbits over a thousand kilometers, decay times go into thousands of years. When you get into geostationary orbits, or high earth orbits, they are effectively infinite. I got really interested in that. In particular, I started to think about spacecraft as artifacts of human culture-really extreme kinds of artifacts of human culture. When it comes to time scales and cultural artifacts, we usually think of things such as pyramids or cave paintings. The time scales associated with cave paintings go back to a few tens of thousands of years or so. Pyramids are between two and three thousand years old. And they are falling apart. When you look at the time scales that satellites exist on, you're talking tens of thousands, hundreds of thousands, millions, billions of years. This is quite dramatic in terms of a human footprint in the cosmos. This is orders of magnitude greater than anything that humans have done before. That's how I got interested in what we might call the human geology of outer space.

**SMUDGE:** We've been intrigued by your desire to go places yourself and place your body and practice at the edge of these events. Whether it's the "black sites" project that had you taking photos related to secret military bases from distant Nevada mountaintops or from the window of a Las Vegas hotel overlooking the airport, you've been drawn to actually going there and feeling it for yourself. This is an important part of what we are trying to illustrate with idea of a "geologic turn," this exposure and visceral response to actual event-ness, or to change or forces. Could you tell us about the way you've designed your practice to actually take you into the world, and if part of the work itself is being exposed to its material realities?

**PAGLEN:** That's a crucial part of what I do. I'm trying to remove as much mediation as I can, and of course, to do that completely is impossible. But if you go to a place, you see all kinds of connections and possibilities and constraints in the materiality of what you are looking at, that you would never be able to see if you didn't go there. This happens over and over again. I think a lot of us have gotten used to getting information or experiencing the world through the information funnels that we have, whether that's reading the news or Twitter feeds. Those very mediated forms close off a lot of things that you would otherwise see if you went to the place where something was happening. For example, I was collaborating with a friend of mine who is journalist. We were looking at different front companies that the CIA had set up to do rendition programs. One of the places we were looking at was a law office in Reno, Nevada. We drove up there and went to the law office, and we realized that there was a former senator who had his west coast office in the same suite as the lesser known law office, which was the home of the front company involved in the rendition program. That was something that you had to be there to realize. You would completely miss that if you just did record searches. Going to the place allows you to make connections that you wouldn't otherwise be able to make. The world is a very complicated place. The more that our understanding of it comes from highly mediated sources, the more complexity gets bracketed out. We don't even realize the extent to which this happens. ■

"In a nonlinear world in which the same basic processes of self-organization take place in the mineral, organic, and cultural spheres, perhaps rocks hold some keys to understanding sedimentary humanity, igneous humanity, and all their mixtures."

—Manuel DeLanda,
*A Thousand Years of Nonlinear History*

image: Elizabeth Ellsworth, Great Salt Lake Desert, 2010

Ilana Halperin

## 23. AUTOBIOGRAPHICAL TRACE FOSSILS

Ancient and technologically advanced modes of production were employed to arrive at these unique objects of pure geology; including traditional copper plate etching, virtual modelling, rapid prototyping and limestone encrustation. Image courtesy Ilana Halperin.

### Physical Geology (A Geological Time Diptych)

Ephemeral islands, petrified raindrops, volcanic dust from Pompeii, a shared birthday with a volcano. Certain geological processes have a way of evoking a more personal response to the idea of geological time. What does a time span of 300,000,000 years actually mean? *Physical Geology* is an ongoing project which explores our desire to make corporeal contact with geological phenomena.

While conducting research as an Alchemy Fellow at the Manchester Museum, in the "oddities drawer" of the geology department I came across a fine collection of lava medallions from Mount Vesuvius, magma pressed between forged steel plates to form an imprint (imagine a waffle iron that uses lava as batter). In the same drawer, a small stone relief sculpture appeared to be carved out of pure white alabaster. In fact, it was a limestone cast created via the same process that forms stalactites in a cave.

In re-visiting these historical geological art processes (lava forging and cave casting), I began to develop ideas for "physical geological art works," art objects formed within a geological, or deep time, context. I went to Saint Nectaire, a small thermal town in the mountains of the Auvergne in France to meet with Eric Papon, whose family founded the Fontaines Petrifiantes, and the art of cave casting, seven generations ago. In a normal limestone cave it takes one hundred years for a stalactite to grow one centimeter, while in the caves of the Fontaines Petrifiantes one centimeter grows in one year. Every year Eric and his assistant facilitate the formation of hundreds of cave casts through a labor-intensive process involving twenty-five meter high casting ladders, carbonate waterfalls and natural rubber moulds which slowly fill with limestone deposits over the course of a year. In 2008 we began to work together in the caves.

The cave cast can be likened to a drawing, a record of incremental change. It is a cousin of the long exposure found in pinhole photography; the gradual accumulation. Though a cave cast serves as a stand-in for deep time, twelve months are not massive in a geological time context, in daily life a lot can happen in a year. Within an arts context, a year is also quite a long duration, a continually occurring process forming a piece of work.

The plan is to make a geological time diptych, new lava medallions, new cave casts, slow time and fast time alongside each other. To date, I have made a new series of cave casts which formed over the course of ten months in the calcifying springs of the Fontaines Petrifiantes. The last casts were cracked open to coincide with my 36th birthday in September 2009. The slow time component of the project is now complete.

Ancient and technologically advanced modes of production were employed to arrive at these unique objects of pure geology; including traditional copper plate etching, virtual modelling, rapid prototyping and limestone encrustation.

In a volcanic prologue to the fast time component, I spent time encountering active lava flows as they formed new land on Big Island, Hawaii. To complete the project, lava-stamping implements are ready and waiting for a molten river to emerge from another volcano, Etna, where the art of lava forging surfaced. To carry out this action, volcanologists have agreed to work together at the edge of the next flow. This may occur in one moment or within a lifetime.

In a volcanic prologue to the fast time component, I spent time encountering active lava flows as they formed new land on Big Island, Hawaii. To complete the project, lava stamping implements are ready and waiting for a molten river to emerge from another volcano – Etna, where the art of lava forging surfaced. To carry out this action, volcanologists have agreed to work together at the edge of the next flow. This may occur in one moment or within a lifetime. Image courtesy Ilana Halperin

### Physical Geology, Geological Intimacy

Within my work, drawing parallels between very personal events, for example when I was born or when my father died with the birth of a volcano, allows for a space to think about our place within the geological time continuum from a more intimate perspective. The next stage of *Physical Geology* extends these links into new geological terrain.

One evening in Edinburgh, someone approached me and said, I have been thinking about your work lately, I came across something that I think you might be very interested in, it's a collection of body stones. *Body stones?* Body stones, gall stones, kidney stones, they are all made of geology. Out of this conversation grew a totally unexpected line of enquiry within my work, the idea that we as humans are also geological agents, we form geology. We are like volcanoes, producing new landmass on a micro scale. We are closer relations to Eyjafjallajokull than previously thought. The boundary between the biological and geological can begin to blur. *Physical Geology (geologic intimacy)* is as an exploration of new landmasses of a cultural, biological and geological nature, from petrifying caves in France and geothermal pools in Iceland to a collection of body stones more animal than mineral, or perhaps somewhere in between.

I have spent time with two collections of body stones, both in Berlin. One is a historical collection of stones from the 1700s, excavated by Johann Gottlieb Walter and Friedrich August Walter, a father and son team of body mineralogists. The second collection is contemporary and belongs to Navena Widulin, the medical museum preparateur. Encountering Navena's collection on display in her laboratory, you would think you were looking at shelves of precious gems and minerals. She began her collection of body stones in the mid-1990s. Now surgeons and pathologists throughout Germany send her stones in the post whenever they extract one from someone living or dead. In the body, each stone is a biological entity, and once out of the body it belongs to the realm of geology.

I had a memory of seeing a cross section of an old pipe in the geology collection of the Hunterian Museum in Glasgow. It's a thick pipe, about a foot in diameter, though you can hardly see through it as a thick geothermal mineral deposit formed inside, like the start of a clogged geological artery.

Building on earlier investigations into the formation of culturally occurring landmass, I could not kick thinking about that pipe. In February 2011, I began working with the Blue Lagoon in Iceland to generate a new series of silica and mineral sculptures which form in geothermal pools. Working alongside Hannes Johannson, the foreman of the Blue Lagoon, we created a new series of geothermally occurring sculptures in the runoff pool from the geothermal power plant, which feeds the Blue Lagoon. The sculptures formed over a period of 2.5 weeks time. They are composed of laser-cut plywood stencils submerged in geothermal pools to naturally encrust in pure silica over the period of formation.

### February 2011. Geothermal Diary. Grindavik, Iceland

Every day, I wake up in the pitch-black night, but it is morning. I go to the industrial waterfall with Hannes. We are surrounded by steam and silver silos. An active geothermal power plant fed by an underground geothermal sea. We are inland. We check our work. How much have you grown today? One line is reeled in, the skeleton is almost encrusted. We turn a second, like basting a roast for Shabbat dinner; one side, flip, then flip again. We climb the scree black lava slope, past red rust and copper green to the upper level of the waterfall. A green rope dangles down into the mist disappearing as if into a well or a cavern. Two ribs over, one line attached to *corporeal mineralogy*, one to *autobiographical trace fossils*. Each one in a new grey skin, like a young rhinoceros or an armadillo, almost translucent, but there is no question this worm skin will become strong and solid. We are in the high temperature pools here, 80 degrees Celsius, where we learned everything forms just as you hope.

The plan is to make a geological time diptych—new lava medallions, new cave casts—slow time and fast time alongside each other. To date, I have made a new series of cave casts that formed over the course of ten months in the calcifying springs of the Fontaines Petrifiantes. The last casts were cracked open to coincide with my 36th birthday in September 2009. The slow time component of the project is now complete. Image courtesy Ilana Halperin

To get to our morning check, we march through crisp new snow, crystalline, sparkling over pipes filled with boiling water, well insulated, plastic. You would never guess from above what courses through the pipeline underfoot. Hot heat that would melt snowfall or skin. We cross lava rock covered in ice, uneven surface. Hannes says you must already have chosen your next step as you take your first. Confidence over surface so as not to fall, falter, fear anything broken or cracked on a rock. We cross the clear warm rainforest dew of the power plant, melting away any last trace of ice or snow. Past the cooling tower, not a tower so much as a massive box with water running over, top to bottom, like a perpetually overflowing bathtub. So much energy is generated here it powers the entire region. (Hard now to picture this, so similar in look to Fukushima, though an entirely different animal.) Eventually, we come to the waterfall, strewn with bones of silica coral, and channels like woven reed baskets encased in calcium.

This is what we do. We form geology. We check each piece, turn it. I can't think of a time when I have been able to watch something with such attention. Memorise the details, compare them the next day, the day after that. Watch the sunset roll in, like clockwork tides that shift each day. A small slice of light added onto to the hours of February winter dark. When I arrived, five hours of day and twilight, now six and counting. ■

Endnotes

The Allende meteorite is the oldest known object on earth, older then the earth itself. I spoke with a paleon-tologist who told me a fossil is the presentation of the moment of death. However *trace* fossils record an action, eating or walking, but not the organism itself.

What is the correlation between a volcanic island, a cave cast and a body stone? Metamorphic rock comes from the Greek to "change form." On top of a mountain, I found fossils from the ocean floor. In the early days of exploration, new landmass was named: New Amsterdam, Heilprin Land, Ferdinandea. How do we locate a body stone on the map? Should we mark the birth and death of an island? A cave cast is a free-floating landmass that can be passed from one person to another. A body stone, a miniature planet travelling through an interior universe. The geological record may need to be re-assembled.

We are autobiographical trace fossils.

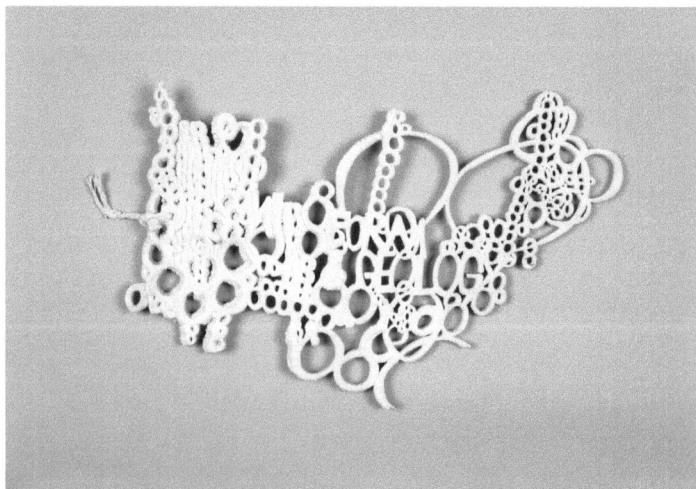

I began to develop ideas for 'physical geological art works' - art objects formed within a geological, or deep time context. The cave cast can be likened to a drawing, a record of incremental change. It is a cousin of the long exposure found in pinhole photography; the gradual accumulation. Though a cave cast serves as a 'stand in' for deep time, twelve months are not massive in a geological time context, in daily life — a lot can happen in a year. Within an arts context, a year is also quite a long duration, a continually occurring process forming a piece of work. Image courtesy Ilana Halperin

Rachel Sussman

## 24. LIVE THROUGH THIS: SURVIVING THE PLEISTOCENE IN SOUTHERN CALIFORNIA

Creosote Bush #0906-3637 (12,000 years old; Mojave Desert, California)

The Oldest Living Things in the World are continuously living organisms 2,000 years old and older. Starting at "year zero" and looking back from there, they help reframe our personal timescale in a shift towards the long term well beyond a single human lifetime. There are a number of characteristics that many of these ancient organisms have in common, such as slow growth rates and the ability to thrive in adverse conditions. Likewise, there are geographic concentrations of these multi-millennials in the Mediterranean and Australia (and it should be noted that there are many parts of the world that have not enjoyed the same dedicated scientific attentions), but one of the largest known geographical concentrations of ancient life anywhere on the planet happens to be in Southern California.

If you ask someone what the oldest trees in United States are, the answer is often the Redwoods. It's an understandable mistake: they are majestic, breathtakingly large, and admirable in both girth and height.  But while the Redwoods hold records for certain superlatives, their cousins to the south, the Giant Sequoias, are older. And even then, the Giant Sequoias are in fact the youngest of the five species in California alone that have surpassed the 2000-year mark.

Some are aware of the Bristlecone pines, which have the Sequoia genus beat out by over a factor of two. The Methuselah tree is the most celebrated, followed narrowly by the infamy of the Prometheus, which would still hold the record as oldest, had it not suffered the indignity

*Mojave Yucca #0311 -P0983 (12,000 years old; Mojave Desert, California)*

of being chopped down in the mid 1960's to retrieve a lost coring bit. There are actually several living individuals proven to be older than the approximately 5000-year-old Methusela, but their exact identities are jealously guarded for their own protection, and rightly so. Certainly no researcher would chop them down today, no matter what the bit. However, visitor "souvenir" taking has proven quite damaging.

But now let's jump another 5,000 years into the past, where we get into a somewhat more complex set of parameters: organisms growing clonally. Sometimes referred to as "self-propagating" or as "vegetative growth," these individuals are able to generate new "clones" of themselves, if you will, as opposed to reproducing sexually. In other words, they create new shoots, stems, roots, etc., without the introduction of outside genetic material. Thus the new growth is genetically identical to, and part of, the original organism; a process that can continue indefinitely, or as long as outside environmental factors allow. It's not that these organisms *can't* reproduce sexually, it just seems that sometimes it's easier to just do it yourself than find a suitable partner.

The approximately 12,000-year-old Creosote Bush and Mojave Yucca sit cordoned off by wire fences on Bureau of Land Management property designated for all-terrain vehicle use. They have remarkable circular structures, pushing slowly outward from a central originating stem. There is no hole to bore that would prove out their remarkable longevity, but rather a slow and steady continuation of self: new stems replace old ones, getting larger by tip toe rather than leaps and bounds. According to a BLM ranger, native populations used the Mojave Yucca's natural Stonehenge formation to take shelter from desert storms.

Let's go back yet another thousand years. What was going on in California 13,000 years ago? For one thing, camels still roamed the area. In fact, camels originated in North America around 100,000 years ago, and made the passage across the Bearing Straight into Asia. (Imagine what the traffic must have looked like.) Mastodons still grazed the hillsides along with other megafauna, their final gambols before decline and extinction. And the still-living 13,000-year-old Palmer's Oak was there to witness it.

Today the Palmer's Oak lives unprotected, if obscurely, within the bounds of the industrial city of Riverside, littered with garbage and the detritus of discarded meth labs. But don't picture a towering tree; it is a scrub oak, or shrub, that is so unremarkable in appearance that it lived undiscovered on private property until a local botanist happened upon it and knew it was something special.

These organisms survived the end of an ice age, lived through megafauna extinctions, and weathered the meteoric rise in human population. How? Why? We know some of these answers on an individual level, but the science of comparative longevity across species is so new it doesn't yet exist. We don't yet know if, or how, this tantalizing longevity might be applied to human lifespans.

But what does it mean from a relational perspective when the organic goes head to head with the geologic? We start talking about longevity and the quotidian in the same breath. Geologic time, deep time—these concepts that are rock-hard in a way that makes them impenetrable to the human condition start to soften and change with the introduction of the organic. The Oldest Living Things image are photographs of the past in the living present, portraits of indivduals that put a face to a name, and forge a personal connection to a timeframe well outside of our temporal comfort zone.

These handfuls of organisms were able to survive, and thrive, through an explicit conditional shift on the planet.

Will they survive us? ■

Palmer's Oak #0311-0514 (13,000 years old, Riverside, California)

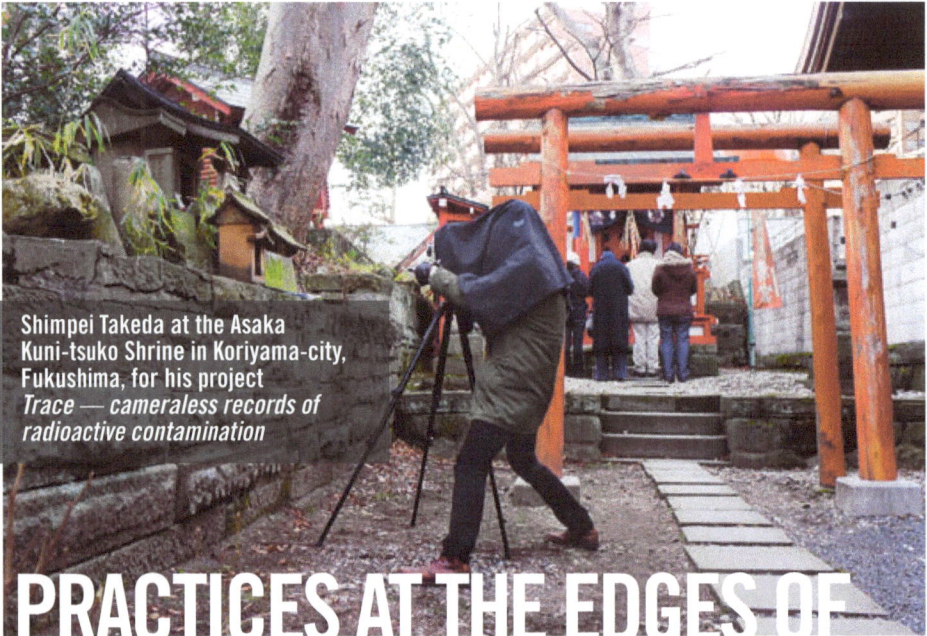

Shimpei Takeda at the Asaka Kuni-tsuko Shrine in Koriyama-city, Fukushima, for his project *Trace — cameraless records of radioactive contamination*

# PRACTICES AT THE EDGES OF
# MATERIAL AND TIME

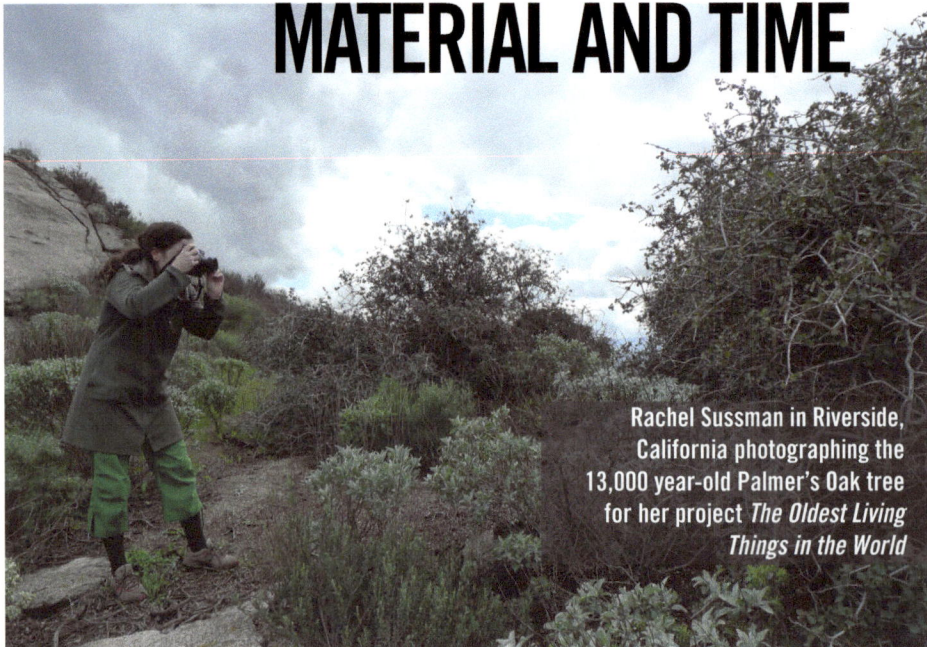

Rachel Sussman in Riverside, California photographing the 13,000 year-old Palmer's Oak tree for her project *The Oldest Living Things in the World*

Shimpei Takeda photo by Keisuke Hiei, 2012, Rachel Sussman photo by Marie Regan, 2011

Matt Baker + John Gordon

## 25. UNCONFORMITIES, SCHISMS AND SUTURES: GEOLOGY AND THE ART OF MYTHOLOGY IN SCOTLAND

> Geopoetics is concerned, fundamentally, with a relationship to
> the earth and with the opening of a world.
>
> —Kenneth White (2003)

James Hutton's classic section at Siccar Point, showing the "unconformity" between steeply dipping Silurian strata and the overlying Devonian (Old Red Sandstone) rocks. Hutton's great insight was that the "unconformity" represented an immense period of "missing" geological time. He showed that the processes shaping the Earth today are the same as those that operated in the past and that the rocks on the surface of the Earth have been recycled many times through erosion, sedimentation on the ocean floors, consolidation, and uplift. Photo: John Gordon

### Opening New Worlds

In the late 18th century, James Hutton's insights opened new worlds, both geological and cultural. For the first time, human identity could be placed against a geological timescale, allowing the emergence of new mythologies. Hutton's theory, underpinned by the concept of geological "unconformity," the interruption in the "rock record" of the natural sequence of rocks, led to a new scientific and rational understanding of geology. That new understanding led to the questioning of creation mythologies through empirical science, creating senses of displacement or "unconformity" among people during the Enlightenment. Humans found their senses of self and their traditional mythologies displaced in relation to, or not conforming to, the notion of deep time.

As the science of Geology set about developing an empirical understanding of the "abyss of time" (as described by John Playfair) and reconstructing Earth history (the "worlds before Adam" in the words of Martin Rudwick), artists wrestled with a different problem: how to build new mythologies that situated humankind's *sense of self* in relationship to this longest of all time-signatures. The different historical responses we present here conclude with the production of "nonconformities" in the landscape by contemporary artists who are re-appropriating geologic science as the material for new mythologies that connect human time to deep time.

In acknowledging this threshold in the history of our relationship to landscape and the "abyss of time," it is useful to reflect on our fascination with the cultural understanding of geomorphology held within the mythologies of aboriginal societies. Perhaps we can begin to accept that the Enlightenment, while bringing new freedoms and possibilities, also left a void to fill: the need for a *creative mythology of self* in relation to the rocks:

> . . . human experience in this difficult Northern place has been built so intimately into the geology and the post-glacial ecology of Scotland that a people and its stones form a single cultural landscape. (Neal Ascherson, 2002)

The Southern Uplands of Scotland form the geological context of this essay. Here, a narrow belt of elevated terrain runs across the country, marking both an ancient suture and the political boundary with England. The feature was formed 425 million years ago when the crustal fragments that would become Scotland collided and fused with one another as a result of the movements of the Earth's tectonic plates. Hutton's Unconformity at Siccar Point lies at the eastern extremity of the Southern Uplands.

In considering the time-span from the Scottish Enlightenment of the mid to late 18th century until the present day, we propose three eras of artistic mythology in landscape: "Tourism of Awe," "Mechanics of Conquest" and "Search for New Unconformity."

### The Tourism of Awe

The Romantic Movement and the quest to experience sublime and picturesque landscapes during the late 18th and early 19th centuries were influenced by the aesthetic values expressed by Thomas Burnet, Joseph Addison, Edmund Burke, William Gilpin, and others. Enlightenment questioning of divine creation as the definitive generator of landscape allowed artists, poets, and writers to re-discover a new awe in the "works" of nature which they contrasted with the works of man. This ability to speculate about unknown but unquestionably awesome creative forces at work in nature presented the challenge to artists to redefine the scale of landscape in relation to mankind—not as something else made by the same benevolent creative hand, but rather, as something "other," something that was literally perilous to contemplate. One such example is the Falls of Clyde (the source of the River Clyde is on the Southern Uplands) where:

> . . . you are struck at once with the aweful scene which suddenly bursts upon your astonished sight. Your organs of perception are hurried along, and partake of the turbulence of the roaring waters. The powers of recollection remain suspended, for a time, by this sudden shock; and it is not till after a considerable time, that you are enabled to contemplate the sublime horrors of this majestic scene. (Thomas Newte [William Thomson], 1788)

"Five Sisters" Oil Shale Bings — Midlothian. These relics of industrial process were the focus of artist John Latham's period as Artist in Residence to the Scottish Development Office in 1976. Latham studied the ecology of the bings in minute detail and proposed their designation as a monument. Photo: Lorne Gill/Scottish Natural Heritage

## The Mechanics of Conquest

With the Industrial Revolution, humans acquired the ability to engineer and shape the landscape, building railways, canals, bridges, and tunnels in a manner unthinkable to the Romantic poets. This ability often turned geology from something inducing awe instead into a series of challenges that could be overcome through human ingenuity and powerful new machinery. This was also the age that gave birth to the first "science fiction." In its incomparable durability and longevity, geology became one of the ultimate contexts against which human achievements and ingenuity could be measured. A romantic age of science offered a human myth of conquering time. Jules Verne claimed Scottish ancestry and spent time touring the country—his science fiction novel *The Underground City* is based in The Trossachs at Aberfoyle. In it, he imagines a community living underground in an abandoned coal mine beneath a great loch and rivers.

In central and southern Scotland this era expressed itself in the landscape through the growing spoil heaps of oil shale and coalmining and the systematic "clearance" of indigenous crofting culture from the land to make way for industrial sheep husbandry for wool production. Artistically, the "Mechanics of Conquest" evolved from the confidence of power into a period of questioning and metaphorical use of the landscape to express concerns about the human condition, as in the poetry of Hugh MacDiarmid, Sorley MacLean, Norman MacCaig and Ian Crichton Smith:

Who possesses this landscape?
The man who bought it or
I who am possessed by it?

—Norman MacCaig, *A Man in Assynt* (1990)

## The Search for New Unconformity: the Southern Uplands

In the most recent era we identify a shift from artists speculating on landscape at one remove to working directly in and with landscape. We see this "Search for New Unconformity" as a strategy to use landscape and geomorphological processes as a framework to question the schisms of specialisation and embrace a process that restructures a mythology of mankind as part of the land.

One of the methods employed by contemporary artists is to creatively re-appropriate the scientific processes of geology as a new mythology inviting humans today to *re-place* themselves in the context of immeasurable time signatures. These new "unconformities" placed deliberately and unceremoniously within the public landscape re-interpret the language of science to re-assert our understanding of ourselves as *part* of a the geologic as a living system, a concept originally embraced by James Hutton in his *Theory of the Earth*. This mirrors the calls to reconnect with the natural world and rediscover a sense of wonder. Ultimately, in the words of the Scottish poet and internationalist, Hugh MacDiarmid:

> We must reconcile ourselves to the stones,
> Not the stones to us.
>
> —Hugh MacDiarmid, *On a Raised Beach* (2002)

## Projects

*Shinglehook* – Matt Baker 2006
Clockwise from top: Overall view looking towards the Meggat Water;
Oak anchors half-buried in the shale; Bronze casts floating on the loch.
Photos: Allan Devlin

### 1. *Shinglehook*, Matt Baker 2006

At one of the highest elevations in the Southern Upland, St. Marys Loch was a significant inspiration to Romantic poets such as Wordsworth, Scott, Hogg and Burns. *Shinglehook* is a "permanent" installation in the landscape that is part time-based experiment and part romantic speculation in futility.

The Meggat Water (river) enters St Marys Loch on the northern shore. The slowing of the river causes the water to release the solid material that it has picked up on its journey down the glen; this deposited material is forming a new spit of land that will eventually divide St. Marys Loch in two. It is estimated that the process will take 10,000 years to complete.

*Shinglehook* is conceived as a "place to wait for geology"—four bronze castings were made from a mould taken from the furthest tip of the growing alluvial land bridge. These metal casts float in a line projecting out from the southern shore towards the Meggat Water; the casts can move laterally but are anchored to the shore via cabling secured by two oak drag anchors. The oak anchors are half buried in the shale shore; free to move with changes to the shape of the land, their form will, ultimately, keep them gripped by the substance of the shale.

The viewer is invited to speculate that, one day, the bronze casts will be enveloped by the arrival of the land bridge across the loch and the mobile structure of *Shinglehook* will become static.

The materials of *Shinglehook* cannot possibly last 10,000 years unaided. Either the project will require the support of people in the future to maintain and renew it, or, it will gradually disintegrate and be dispersed in the landscape. In recent years the work was saved by a passing visitor who extinguished a fire in one of the oak anchors (the local authority then paid for the timber to be repaired) and the local sailing club has begun to monitor the condition of the bronze floats and report any issues to the artist.

*Striding Arches*, Andy Goldsworthy 2009. One of three self-supporting 4m high arches in the landscape. A fourth grows from a traditional stone byre. Photo: Mike Bolam

### 2. *Striding Arches*, **Andy Goldsworthy 2009**

Andy Goldsworthy is an internationally renowned artist working directly in the landscape —his home and studio are in the Southern Uplands of Scotland. He has made many small interventions in the local landscape, but in 2009 completed his first major work for the area, *Striding Arches*. Crucial to Goldsworthy's selection of sites was his stipulation that no matter which arch you find yourself at, you should always be able to see the other two: the three arches are linked together by sightlines.

The overall work comprises a small building for community events and a challenging hill-walk around all the arches.

*Striding Arches* is part of a worldwide project of Goldsworthy's that uses arches made from Dumfriesshire sandstone to highlight the Scottish diaspora. He has constructed arches in Canada, New Zealand and USA, echoing the travels of emigrating Scots over the last 200 years or so, and of the sandstone carried as ballast by timber ships making the return voyage to North America:

> The stone is a potent symbol of the Scots who went abroad of the tremendous upheaval they made, or were forced to make, when they left Scotland... I would hope that collectively these arches are a celebration and monument to the Scottish people and the travels they have made, and that they will act as a connection between those who have left and those who have stayed here.
>
> —Andy Goldsworthy

*Erratic* – Matt Baker 2008.
Bronze chain, pulling handle and fixing on carved granite boulder weighing approximately 100 lbs.
Photos: Matt Baker

### 3. *Erratic*, Matt Baker 2008

*Erratic* is one of 5 time-based works by the artist, "permanently" installed in the Southern Uplands on the slopes of Cairnsmore of Fleet. The area is a National Nature Reserve (maintained by Scottish Natural Heritage) and is particularly noted for the clear evidence of geological processes, including tectonic plate suturing, later granitic intrusions, and extensive glaciation.

Matt Baker collaborated with locally-based poet, Mary Smith, and each work comprises a poem and a physical installation. Sited in one of the most remote and beautiful parts of the landscape, *Erratic* is on a journey. The background information about the project invites anyone finding *Erratic* to move the sculpture to a location of their choosing; the bronze pulling-handle can be drawn from the carved granite boulder and used to physically drag the stone over the landscape. *Erratic* was originally placed approximately 3 miles from the nearest road and is situated in approximately 10,000 acres of open country. ∎

***Erratic*** * (Mary Smith, 2007)

A'm no keen oan bein cried an 'erratic':
soons as though A staggert here,
stottin fu. Tellin ye, it wisnae me:
a big glacier did it –
an then it meltet away.

A wisnae daein nothin wrang whin
it gied me a shove, wheeched me up,
whirled me aroon lik A wis
nae mair wechtie nor a balloon.
Next meenit, heid ower heels,
breengin doon thon hill.

* *An erratic is a boulder transported and deposited by a glacier. They are useful indicators of patterns of former ice flow.*

References

Ascherson, Neal. 2002. *Stone Voices.* London: Granta.

Goldsworthy, Andy. 2009. *Striding Arches*: http://www.stridingarches.com

Hutton, J. 1788. "Theory of the Earth; or an investigation of the laws observable in the composition, dissolution, and restoration of land upon the globe," *Transactions of the Royal Society of Edinburgh*, Vol. 1: 209-304.

MacCaig, Norman. 1990. *Collected Poems.* London: Chatto & Windus.

MacDiarmid, Hugh. 2002. *Selected Poems.* Manchester, UK: Carcanet.

Newte, T. [William Thomson]. 1788. *A Tour in England and Scotland, in 1785.* London: G.G.J and J. Robinson.

Playfair, J. 1805. "Biographical account of the late Dr James Hutton, F.R.S. Edin.," *Transactions of the Royal Society of Edinburgh*, Vol. 5: 39-99.

Rudwick, M.J.S. 2008. *Worlds Before Adam. The Reconstruction of Geohistory in the Age of Reform.* Chicago: University of Chicago Press.

White, K. 2003. *Geopoetics: Place, Culture, World.* Glasgow: Alba Editions.

Chris Rose

## 26. TIME AND THE BREATHING CITY

Image: Chris Rose, Vertical section of marine rock formation, Portland Stone Quarry, 'Portland Bill' Dorset UK.

During the early research phase of a large Arts-Sciences-Design collaboration in the UK, I joined a site visit to the Portland stone quarry on the south coast of the UK, part of the area designated as the "Jurassic Coastline" World Heritage Site. I was part of a small team comprised of meteorologist Dr. Janet Barlow, composer and sound expert Holger Zschenderlein, and myself as artist and designer. We were in the midst of exploring concepts of time, material evidence, data, and embodied cognition/experience especially connected with complex systems in the atmosphere and our understanding of them.

We started our process by visiting the Portland Stone quarry (Portland Bill, Dorset UK). This site presents a kind of "time interface" in two entirely different ways: 1) the collision of a rock formation of equatorial marine origin; Portland Bill itself as part of an "object" traveling from the equator eventually to what is now the south coast of England and embedding in the sedimentary rocks of that place. Here the differences in material properties along the collision zone altered the local erosion behavior and continues to modify the coastline. Places are accessible where a cross section of the collision zone can be seen; and 2) places where a vertical slice is visible through the integral, layered record within the Portland stone itself of alternating periods of forestation, shoreline formation, and subsequent reforestation forming a kind of "grammar" of observable repetition.

This material history has been conveyed intact over planetary distances to the Portland area. The dominant perception for me in this remarkable conjunction of phenomena became now inverted. The rock I was standing inside of and adjacent to was now "process," rather than as it had been before I arrived, namely, an area of England on the map, "a vague place," in the abstract. I had used a map to travel to this "place." The abstraction carried in my mind was now confounded by the paradoxical hardness of material; a material that had consolidated nuances of alteration over planetary time; literally a parallel universe to that of which I was presently conscious by virtue of the discontinuity of its, and my own, time scales. The hardness was yet again dissolved in sensory terms because the processes embodied were those of flow, of weather, of tropics and temperate change, of coastline, forest, plant and marine living process all within the illustrative "rock." It was as solid as it gets yet was all about "flow." A beach environment could be seen overlaying a collapsed fossilized forest from a previous era. In turn at a later time, the fossilized beach had been overlaid by a different ecology, and so it was repeated, both up towards the visible (present) surface, and below, further than could be seen.

This real sense of the inadequacies of the abstract created for me a consequent sense of the existence of an entirely different "outcome space" in my own perception. This is to say that we need not be confounded by apparent contradiction in our knowledge. Paradox does not independently exist in the world. Paradox is more accurately thought of as a description of a subjective "felt" experience signifying a mismatch between what appears to us and what we think we know. There is no such thing as paradox in the world. Paradoxes exist only in terms of the relational properties. And it is within the relational properties that our struggle for meaning occurs. Standing in this place, I experienced a perception of the materiality of time which felt simultaneously incomprehensible yet real and incontrovertible. It was as if I felt myself pushed in the direction of a realisation without being able to understand it.

Image: Chris Rose, Sedimentary rock folded up by pressure from collision of Portland Rock mass travelling from the equator. Stair Hole, Dorset.

**Breathing CIty Installation, "Festival of Science," London 2010**

Four years after the rock visit, Holger, Janet and myself, joined now by Patrick Letschka, another designer, were concentrating on a specific installation for the Royal Society of Science on London's South Bank. The installation linked aspects of urban weather effects, data visualisation and the inherent unpredictability of complex systems, coupled with the thought that everyone is "connected" to weather.

In the period since our initial visit to the Portland rock, we had been discussing ideas about the "materiality of time" and we were deliberately noticing anything that was connected with them. We worked with soundscape design, influences of "big data" and relationships between the complexities and depth of our evolving, embodied cognition, and how these provide a kind of connective tissue in the language of arts and science collaboration. Themes of time, materiality, changing forms of energy and mass, and how evocation via everyday embodied cognition of complex systems (in this case, the atmosphere, energy dissipation, ice, and sound) can be empowered within an encounter space. This was our "design problem."

We installed a two-ton block of ice within an architectural interior space. It provided a complex anchor for a surprising range of human perceptions and emotions. We discovered that the ice installation evoked material and process relationships analogous to those I experienced while standing at the rock face. A two-ton block of ice within an architectural interior space provides a complex anchor for a surprising range of human perceptions and emotions, engaging, as it does, each of the senses. It was solid, heavy, somewhat hostile, yet ephemeral. It was affecting and un-missable. It chilled the air. As visitors, we were confronted by a solid heavy mass that we knew was transient. Where does it go? It contains a huge amount of embodied energy (it took three weeks to freeze) and many consequences of the gradual transfer of this energy are complex, unpredictable and surprising. Material changes occur and re-occur. Light interacts with the material in a changing manner.

All of this presents the potential for highly engaging relationships with groups of viewers (which included specialists, families, and the general public) who were "drawn out" by their experiences and had things to say about "it" to each other. The dynamic at play here is a perceptual drama revolving around the tensions between expectation, anticipation, experience and narrative. We know little but experience much. ■

See cjvrose.com

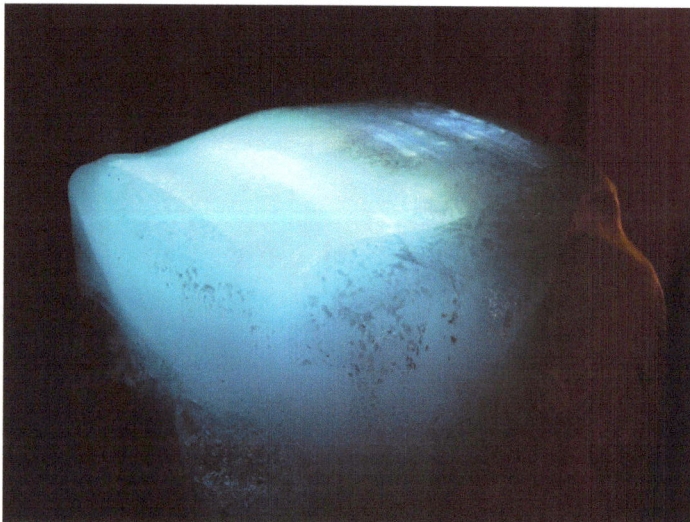

Image: Chris Rose, Ice-Sound-Video Installation by the Breathing City Group, London South Bank 'Festival of Science 2010, titled 'Ice-Traffic'

Etienne Turpin

## 27. ROBERT SMITHSON'S ABSTRACT GEOLOGY: REVISITING THE PREMONITORY POLITICS OF THE TRIASSIC

> A basic problem is linked to the very idea of philosophy: how to get out of the human situation.
> —Georges Bataille

> I'm interested in the politics of the Triassic period.
> —Robert Smithson

Photograph of "remote times," courtesy of the author.

### Politics, or "A Gnaw That is Ratting"

In response to an artist symposium question regarding the "deepening political crisis in America," Robert Smithson published a piece in Artforum, September 1970, titled "Art and the Political Whirlpool or the Politics of Disgust." In it, he said:

> My 'position' is one of sinking into an awareness of global squalor and futility. The rat of politics always gnaws at the cheese of art. The trap is set. If there's an original curse, then politics has something to do with it. Direct political action becomes a matter of trying to pick poison out of boiling stew. The pain of this experience accelerates the need for more and more actions.[1]

Later that same year, when asked about his political stance by Philip Leider, Smithson remarked, "I'm interested in the politics of the Triassic period." In an earlier essay, "The Crystal Land" (1966), Smithson suggested some of the sedimentary compositions at stake in a geo-artistic politics:

Brian H. Mason, in his fascinating booklet, Trap Rock Minerals of New Jersey, speaks of the "Triassic sedimentary rocks of the Neward series," which are related to those of the Palisades.

---

1 Robert Smithson, *Robert Smithson: The Collected Writings*, ed. Jack Flam (Berkeley: University of California Press, 1996), 134. This collection will henceforth be referred to as RSCW and cited parenthetically by page number.

In these rocks one might find:

> actinolite, albite, allanite, analcime, apatite, anhydrtie, apophyllite, aurichalcite, aximite, azurite, babingtonite, bornite, barite, calcite, chabzie, chalcocite, chalcophyrite, cholo- rite, chrysocolia, copper, covellite, cuperite, datolite, epidote, galena, glauberite, goethite, gmelinite, greenockite, gypsum, hematite, heaulandite, hornlende, laumontite, malachite, mesolite, natrolite, opal, orpiment, orthoclase, pectolite, prehnite, pumpellyite, pyrite, pyrolusite, quartz, scolecite, siderite, silver, sphalerite, sphene, stevensite, stilbite, stilpnomelane, talc, thaumasite, thomsonite, tourmaline, ulexite.' (RSCW 7-8)

The inorganic thrust of Smithson's political refusal is clear. Yet, despite the possible read- ings of his position as a de-politicized view of sedimentation and history (and, by implication, human agency), Smithson's commitment to the politics of the Triassic period aligns with his broader geo-/cosmo-political project of an "abstract geology." And it allows us to see that proj- ect's relevance to the geological turn in cultural awareness and to the era of the Anthropocene.

## Abstract Geology

Several works are indispensable for understanding the "abstract geology" that Smithson develops in his "sculpture" and his writing during the late 1960s. In "A Sedimentation of the Mind: Earth Projects" (originally published in Artforum, September 1968), Smithson's contends, "One's mind and the earth are in a constant state of erosion, mental rivers wear away abstract banks, brain waves undermine cliffs of thought, ideas decompose into stones of unknowing, and conceptual crystallizations break apart into deposits of gritty reason. Vast moving faculties occur in this geological miasma, and they move in the most physical way. This movement seems motionless, yet it crushes the landscape of logic under glacial reveries" (RSCW 100).

The cognitive processes are not, in this description, merely metaphorically connected to the material geological tendencies. Smithson's emphasis suggests that there is a contiguous, rather than comparative, relation between mind and matter. After discussing his affections for construction equipment, Smithson goes on to suggest, "With such equipment construction takes on the look of destruction; perhaps that's why certain architects hate bulldozers and steam shovels. They seem to turn the terrain into unfinished cities of organized wreckage." (RSCW 101) For Smithson, "The actual disruption of the earth's crust is at times very compelling, and seems to confirm Heraclitus's Fragment 124, 'The most beautiful world is like a heap of rubble tossed down in confusion'" (RSCW 102).

Later in the same essay Smithson again refuses metaphor and states,

> "The strata of the Earth is a jumbled museum. Embedded in the sediment is a text that contains limits and boundaries which evade the rational order, and social structures which confine art. In order to read the rocks we must become conscious of geologic time, and of the layers of prehistoric material that is entombed in the Earth's crust. When one scans the ruined sites of pre-history one sees a heap of wrecked maps that upsets our present art historical limits." (RSCW 110)

In order to escape the confines of the restricted economy of art history, Smithson thus suggests reading another form of text, that of the geological sedimentation that is culture. As Georges Bataille warns, "Sentences will be confined to museums if the emptiness in writing persists." [2]

If the museum is the space where the iconic shapes and images of culture are confined to a particular, historical, and socially-conservative articulation, Smithson's practice attempts to break this space open and bend the discourse of art into strange contours befitting the improbable circumstance we call existence. Smithson confronts attempts to contain art work within the boundaries of criticism. Writing with particular force against the critic Michael Fried, Smithson asserts:

> "Most critics cannot endure the suspension of boundaries between what Ehrenzweig calls the 'self and the non-self.' [...] The bins or containers of my Non-Sites gather in the fragments that are experienced in the physical abyss of raw matter. The tools of technology become a part of the Earth's geology as they sink back into their original state. Machines like dinosaurs must return to dust or rust. One might say a 'de-architecturing' takes place before the artist sets his limits outside the studio or the room." (RSCW 103)

Photograph of "remote times," courtesy of the author.

---

2 Georges Bataille, *On Nietzsche*, translated by Bruce Boone, introduction by Sylvere Lotringer (St. Paul, Minnesota: Paragon House, 1992), 7.

"De-architecturing" does not merely reiterate artistic freedom against the imagined confines of the museum. Instead, it is a confrontation with the artists' own presumptions about making and the space of production. It is also an abstract geologic practice, which examines and selects the foundational elements upon which these architectures of confinement, both in terms of actual space and potential capacity, rest.

### Triassic Apprentice

Smithson is committed to annihilating any simple historical narrative that would allow the values and morality of Man to infect analysis or practice. Smithson's politics of disgust is entirely related to the problems of history and progress, whether articulated in terms of social and political life, aesthetic refinement, or philosophical inquiry. Perhaps the most striking statement on these points is to be found in Smithson's "travelogue" essay "The Domain of the Great Bear." In it, he recounts his visit to the Hayden Planetarium (originally published in *Art Voices*, Fall, 1966). In the final section, "Illustrations of Catastrophe and Remote Times," Smithson describes an especially convincing mise-en-abyme:

> The problem of the "human figure" vanishes from these illustrated infinities and prehistoric cataclysms. Time is deranged. Oceans become puddles, monumental pillars of magma rise from the dark depths of a cracking world. Disasters of all kinds flood the mind at the speed of light. Anthropomorphic concerns are extinct in this vortex of disposable universes. A bewildered "dinosaur" and displaced "bears" are trapped in amazing time dislocations. "Nature" is simulated and turned into "hand-painted" photographs of the extreme past or future. Vast monuments of total annihilation are pictured over boundless abysses or seen from dizzying heights. This is a bad-boy's dream of obliteration, where galaxies are smashed like toys. Globes of "anti-matter" collide with "proto-matter," billions and billions of fragments speed into the deadly chasms of space. Destruction builds on destruction; forming sheets of burning ice, violet and green, it all falls off into infinite pools of dust. A landslide of diamonds plunges into a polar crevasse of boundless dimension. History no longer exists. (RSCW 33)

The themes in this passage hang from the myth of human progress like corpses from an open wagon, clumsily winding its way to an open grave: cataclysm, derangement, monumentality, disposability, dislocation, bewilderment, obliteration, destruction. They refuse to perpetuate the interminable denial of the ambiguity and absurdity of human existence within the geological and cosmological continua. Here we see Smithson anticipating the Anthropocene most clearly: to think the human divorced from assumptions of purpose.

Photograph of "ancestral catastrophe," courtesy of the author.

Smithson's concern with the entropic decline witnessed by geological time can be clearly read through many works from the 1960s, including his early collage *Untitled (Venus with Reptiles)*, 1963. The Venus collage includes fourteen separate rough cut outs of images of reptiles oriented with more or less interest around a centerfold nude figure, reminiscent of a 1960s Playboy model, whose calm yet depthless repose and accentuated breasts suggest a contemporary Venus under the gaze of the agents of geological time as opposed to her usual cherubs, or, perhaps, a contemporary Olympia, now herself enslaved by the reptilian messengers of deep time. Her sanctity, whether as aesthetic figure or popular model, is threatened by a time scale that far exceeds the human. If we accept the geological referent suggested by the reptilian frame, it is clear that Smithson's preference for Triassic politics is an attempt to break with the limited horizon of European aesthetics and its concrescence in the form of Modernism criticism (i.e. Clement Greenberg and Michael Fried), and the separation of art and criticism implied by this tradition.

But, these reptilian figures, while popularly associated with the legacy of the Jurassic period, also witnessed events corresponding to the politics of the earlier Triassic age (250-210 million years ago). They saw the slow but decisive break-up of Pangaea into the two supercontinents, Laurasia and Gondwana, evidence that any form of stable unity is a fiction undone by the viscous earth. And they observed the morphological emergence of ceratitida, the order of nearly all ammonoid cephalopod genera, whose planispiral shells suggest a coiling figuration that would later be rescaled in Smithson's most well-known earthwork.

Politics can be understood as a confrontation among forces that permit, through various means, the negotiated expression of form. Within that understanding, the testimony of the Triassic in the sedimentary record is a reptilian refusal of the Modernist narrative of artistic medium-specific progress. It is also a demand for artists and thinkers today to confront, through inhabitation and apprenticeship, the disorienting material spiral of mineralization that confronts the myth of human progress in Anthropocene. ∎

"When I was invited to create a work for the New Orleans Museum of Art, I immediately thought of these "edges" where the man-made meets the organic, where today meets yesterday, ten thousand years ago, and tomorrow, and where solid meets liquid meets air."
—Katie Holten

image: Katie Holten, *Study #70 (Bayou Chevreau, Louisiana)*, 2012

Anne Reeve

## 28. JARROD BECK: GEOLOGIC ANXIETY

Jarrod Beck, *Disruption Regime*, 2010–. Cast Plaster on 5 acres, Outside of Terlingua, Texas

It's a straightforward proposition, but one just as often overlooked: what's here is here now, only because of how it was then, and it won't be this way forever. It seems we are uniquely able to appreciate this when the terrain around us changes, when a storm or an interruption (a quake, a tidal wave) jostles our sense of scale and proportion and their accompanying variables shift. A body shrinks and quiets against an unspeakably broad landscape, rain cracks a bright day into pounding blackness.

Yet, how do we measure ourselves against slower, monumental shifts that carve the land around us? What "geologic anxiety" might manifest from our sense of being in a continual, if imperceptible, state of flux?

On five acres in the far West Texas desert outpost of Terlingua, artist Jarrod Beck is carefully enacting a "Disruption Regime" to explore this question. In local intraterrestrial parlance, this "disruption" refers to overgrazing that permanently changes the structure of the ground plane, eroding the root system and in turn the ground's ability to withstand summer rainstorms. These Texas storms then become floods, sending the land on the move and rendering it permanently hostile to many forms of growth. Beck appropriates the term for an experimental "ground-drawing," which he views as a test in marking human action against the top strata of the geology it occupies.

In his own words:

*I make the spaces I make in order to understand change on my own time, in my own parameters. To interact with a landscape in a way that's appropriate to the speck of my existence on the geologic scale. The ground drawing is being made over time, in layers; first I burn mesquite branches in shallow trenches, scorching long lines across the property. Over this under drawing, I place cast-plaster limbs—bright white dashes visible from a distance. When laid on the ground these plasters also serve as catchments, small places for rainwater to pool, collect seeds, and attract animals. My pace in the desert is slow and methodical, but that pace is integral to the development of the drawing. No sudden moves; on a geologic scale of time the project is sudden enough. There have been moments when I have suddenly realized that I need to get in the car and drive away. Although this has something to do with the heat, it's also about knowing that all of these hours of work cannot affect time or landscape. This isn't self-doubt, but more like a "ground-truth", pushing back against any preconceptions I might have brought to the site or tried to impose upon it.*

Jarrod Beck, *Ground-drawing for Capture,* 2010, Cast Plaster, Second Floor Gallery, Marfa, Texas

This approach reflects a foundational tenet of Beck's broader practice: that no one object or installation embodies a finality or closure. For Beck, previous work becomes matter upon which new work feeds and regenerates. Any installation gives us only a momentary glimpse into this larger continuum and its expansive cycle of construction and destruction. Materials are broken down and built back up again (those for "Disruption Regime" have been carted from as far away as Albuquerque, Marfa, Houston, and New York), new relationships are tested, abandoned, or expanded upon. In Terlingua, they come together to hypothesize disruption in the landscape, not only as an environmental force, ideally one that might sustain life rather than trample it, but also as a deliberate artistic proposition whose process is absorbed into its form.

In thinking of Beck's project and a "geologic anxiety," this author remembers a school trip in Mesa Verde National Park, where the four corners of Utah, Colorado, Arizona, and New Mexico intersect. The class was invited to participate in an ongoing archeological dig, each member assigned to a small sandbox that had been erected in an attempt to measure out an immeasurable desert. The actions required were simple enough; dig, be careful, find what you can. It was an offer to carve out an area and focus in, to build an intimacy with a small plot amidst an unapproachably vast expanse. In Terlingua, Beck's parcel is an open plain of shifting parameters.

Jarrod Beck, *Disruption Regime*, 2010– , Cast Plaster on 5 acres, Outside of Terlingua, Texas

His work acknowledges both its present and past, the natural and physical work that built the ground and the layers underneath it. *Disruption Regime* becomes part of a ground-narrative hardened and softened over time, every so often imbued with treasure. ∎

*Project supported by Inde/Jacobs Gallery, Marfa, TX.

Victoria Sambunaris

## **29.** THE BORDER PROJECT

Victoria Sambunaris, *Untitled (Border fence), Near Naco, AZ, 2010*

Victoria Sambunaris, *Untitled (Santa Elena Canyon), Big Bend National Park, 2010*

Victoria Sambunaris, *Untitled (Border view south with grasslands), Hereford, AZ, 2010*

The US/Mexico border spans approximately 2000 miles from the Gulf of Mexico to the Pacific Ocean. This southern border weaves its way through diverse terrains and waterways: rivers, mountains, valleys, grasslands, refuges, reservoirs, reservations, parks, forest, canyons, deserts, dunes, ranches, towns, and farms. The extreme physical diversity of the landscape along the border fluctuates between dense urban sprawl compressed along the dividing fence where one can practically see into a person's home across the border to remote uncultivated desert areas with ancient geological formations where one may not see a soul for days.

An imposing physical barrier creates the division between the two countries. The adjacent landscape on the US side is stripped away to accommodate access for Border Patrol and other federal law enforcement agents. From afar, it appears to resemble an extended landing strip or perhaps an environmental art piece. But at ground level, the conspicuous steel fence is a physical reality.

In October of 2009, I departed from New York venturing south to the border area. My initial intent was to arrive at Big Bend National Park, a place of biological diversity and 500 million years of geologic history, to begin my exploration of the border landscape. My assumption that Big Bend was a haven and absent of political implications was proven wrong when it was revealed that since September 11, this area of the border was closed. No longer could one take a small boat across the river to visit the towns of Boquillas or Santa Elena, Mexico to convene with the locals and sample local cuisine. My encounter with the locals shifted my preconceived conceptual notions to a reified construct and the political limitations of access that I would encounter were initially revealed.

The thalweg or deepest channel of the Rio Grande forms the international border and that boundary is enforced to the detriment of the local people who once thrived on tourism. Regardless, I had the opportunity to speak to the Mexican locals who ignored the restrictions and ventured across to sing traditional ranchera tunes or place their handmade crafts for sale to tourists. ■

Victoria Sambunaris, *Untitled (Farm with workers), Jacumba, CA*, 2010

Victoria Sambunaris, *Untitled (Man on horse), Big Bend National Park*, 2009

Wade Kavanaugh + Stephen B. Nguyen

## 30. 473 INCHES AT 60 FRAMES PER SECOND

all images courtesy the artists

### The Soap Factory, Minneapolis, MN | 2006

*473 Inches at 60 Frames per Second* was a site-specific sculptural installation with reference to the Wisconsin Glacial Episode, a glacial period that left Minnesota and parts of Wisconsin covered with numerous lakes and rivers. At the Soap Factory in Minneapolis, MN, Wade Kavanaugh and Stephen B. Nguyen created a two-part artwork that consisted of a glacial ice sheet made of textured white kraft paper and a time lapse video. The two works were physically and temporally downscaled to human scale. The sculpture was created in a way that the final stop-motion documentation read as if the glacier was advancing over a stone canyon made of brown kraft paper. The time lapse video shows the paper glacier advancing at a rate of 473 inches at 60 frames per second, a metric that shifts physical scale and time, from glacial to human. ■

Katie Holten

## 31. NOTHING FROM NOTHING[1]

Katie Holten, *Study #240 (Bayou Portage, Louisiana)*, 2012

These images are from *Drawn to the Edge*, my solo exhibition at the New Orleans Museum of Art (June 15 to September 9, 2012). In many ways the work is an attempt to capture time—to explore how today's New Orleans is profoundly connected to its past and to consider how quickly we seem to be running out of time.

During a six-week residency this spring at A Studio in the Woods (ASITW), which is located on a bend of the Mississippi River just east of the city center, I looked at the history of the place, focusing my research on the city and its relationship with the river, through both the slow processes of geologic time and the rapid changes of the 20th and 21st centuries. I quickly came to see that New Orleans' history is fundamentally connected to its underlying geology, more so than any other place I've ever been. Five thousand years ago, there was no land where New Orleans is today – it was all open water. Slowly over the millennia, the alluvial plain of southern Louisiana was created as the Mississippi River dropped sediment along its meandering path to the Gulf of Mexico.

I discovered Harold Fisks's *Geological Investigation of the Alluvial Valley of the Lower Mississippi River* in which he charts the course of the river. These beautiful maps show the river as a living thing—changing course, snake-like, over the centuries. This powerful force created most of the land south of Baton Rouge.

Today, the lower Mississippi River is hidden behind concrete levees and locked into position, unable to follow its natural course. The bayous and channels within the city are mostly covered with cement and kept buried underground. I went in search of the invisible river and was lucky

---

1 From Lucretius, *On the Nature of Things*, c. 50 BCE. A copy of the book is included in the exhibition.

to have Monique Verdin, a Houma native and photographer, as my guide.

She told me that the understanding of her community's vulnerable situation has greatly deepened since Katrina. I heard this from many other locals. It is now common knowledge that the city is intrinsically connected to its geological past and muddy foundations. In the complex man-made system that is New Orleans it has become clear that we are both separate from and enmeshed within the natural world. The physical environment inherently limits our technological activities, yet we also possess agency and a capacity to modify our surroundings. We are becoming indistinguishable from "nature."

I met locals who work with land and water and I went on expeditions to places like Cocodrie and Venice, Louisiana. I saw the extent of the problems inherent in the landscapes there, which are literally disappearing. Much of this is due to the activities of oil and gas prospectors. Since the 1920's, they have cut thousands of miles of channels through the wetlands, allowing salt water in. As the salt water spreads, it kills the cypress ecosystem, leading to erosion on a vast scale. I kept finding myself standing at the edge of the land, looking at where the water and earth touch, seeing water lap over now obsolete man-made surfaces. I felt a constant, palpable sense of foreboding.

Seeing from above was essential. I took hundreds of aerial photographs during a breathtaking tour in Charlie Hammond's little Cessna.[2] To follow our flight using Google Earth: zoom in on New Orleans, then fly south-west to Houma, veer south-east to Wonder Lake where you can hover over Exxon Company Road and notice all the straight lines cutting through the wetlands, then follow Montegut Road south to Little Caillou Road and then Cocodrie -- the end of the road. Continue flying south and out into the Gulf of Mexico, over some oil and gas platforms (which you won't see on Google Earth as none are visible—all strategically erased), and over Isles Dernieres Barrier Islands. Flying along the ragged edges of this country, it's easy to spot

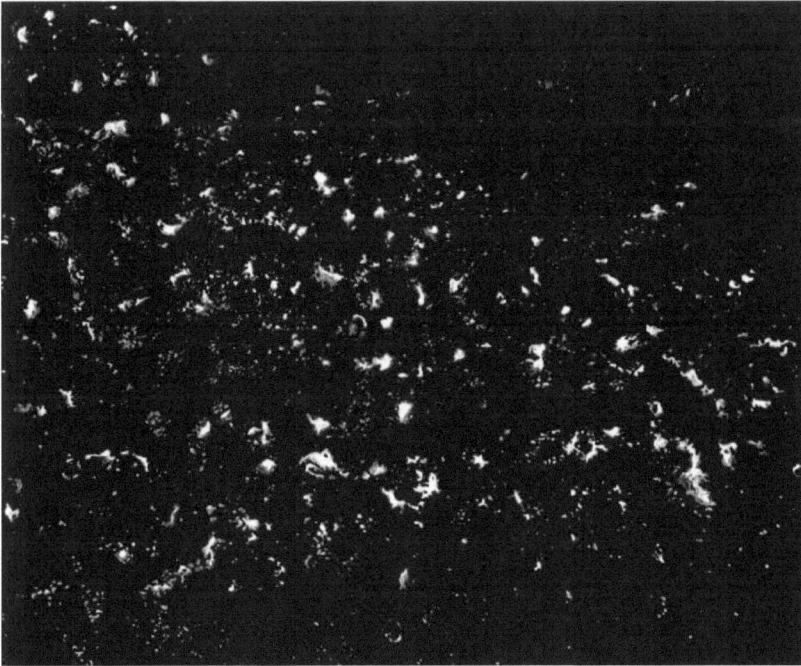

Katie Holten, *Constellations (maps of Louisiana oil and gas wells)*, 2012
chalk from the Cretaceous era on canvas, 10 x 12 feet

2 Katie is currently working on new drawings for a map that will be included in Rebecca Solnit and Rebecca Snedeker's *Unfathomable City: A New Orleans Atlas*, which will be published by University of California Press in Fall 2013.

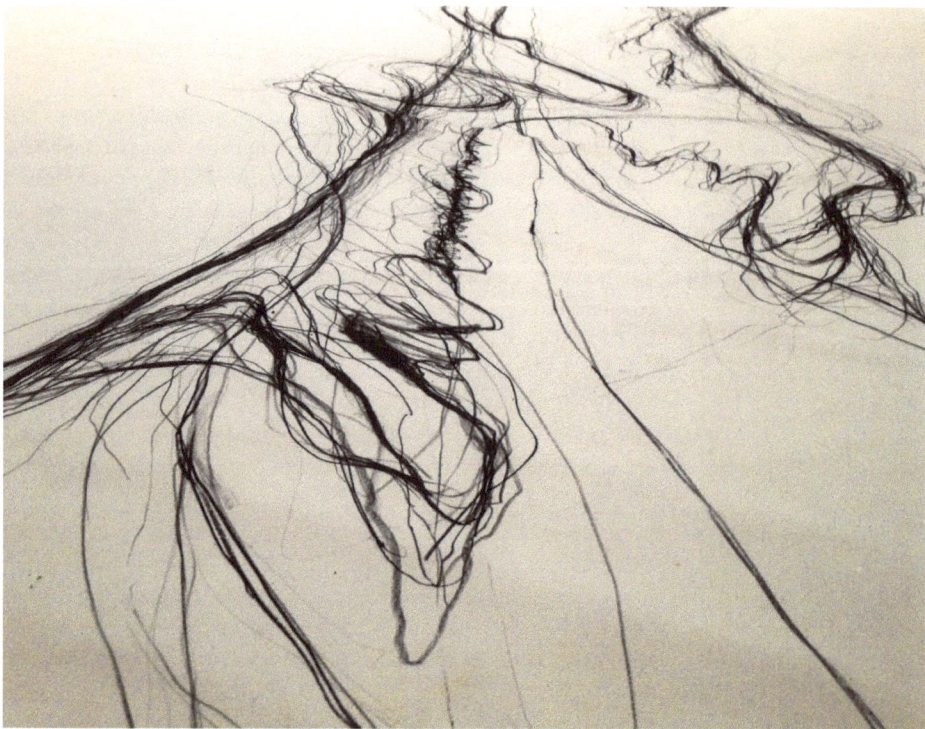

Katie Holten, *Verso: Found Islands (Pine Island)*, 2012. Graphite and charcoal on canvas, 12 x 36 feet (detail)

man-made channels: all the straight lines. Riddled with these, as well as natural channels, the land resembles a piece of lace, or a disintegrating, moth-eaten cloth.

When I was invited to create a work for the New Orleans Museum of Art, I immediately thought of these "edges" where the man-made meets the organic, where today meets yesterday, ten thousand years ago, and tomorrow, and where solid meets liquid meets air. The exhibition title, *Drawn to the Edge,* also alludes to some of my other preoccupations: the exploration of drawing in all its forms, the potential to draw an object while wondering what an object is anyway, and, perhaps most poignantly, the examination of our relationship with the natural world in this age of the Anthropocene, when so many species are drawn to the edge of extinction.

Sculptural in their scale, my drawings for NOMA are all double-sided, made on canvas. Ten and twelve feet tall, they range from sixteen to thirty-six feet long. I used simple materials to draw—graphite, charcoal, chalk, black oil stick, and sediment that I collected from the banks of the Mississippi River.

*Found Islands* depicts, on one side, a charcoal line drawing of Pine Island, which existed 4,500 years ago where New Orleans is today, while the other side features a drawing, made using a light sediment wash, of a contemporary island in Creole Bay formed by the combined processes of man-made events, sediment accumulation, and encroaching salt water.

The narrative within *Drawn to the Edge* is important and the titles help locate the viewer. For example, one drawing looks like a night sky, but the title is *Constellations (maps of Louisiana oil and gas wells)*, so you realize that each of the many thousands of little dots is in fact a well—the seemingly cosmic turns out to be a human-made manifestation of the underlying geology.[3] I made the drawing using chalk from the Cretaceous era. I collected the chalk during

---

3 *Constellations* (maps of Louisiana oil and gas wells) is based on a "Petroleum Industry of Louisiana map" made by Jakob Rosenzweig for a film about Monique Verdin's life titled *My Louisiana Love* (2012).

an expedition to western Kansas when I walked along the former ocean floor of the Inland Sea. Kansas is not a place widely regarded as being intrinsically linked to southern Louisiana. But water from there joins water from about fifty percent of the US interior to flow into the Mississippi River. Sediment from places as far away as Pennsylvania and Montana finds its way downriver. In the spring of 2011, a full year before I ended up in New Orleans, I was on a residency at the Salina Art Center researching water and its relationship with the city of Salina, Kansas. I met with locals such as Stan Cox, a biologist at The Land Institute, who were quick to point out that all the water and runoff from the fields there in Kansas would run into the Mississippi and contribute to the "dead zone" in the Gulf. I didn't know it at the time, but it was all connected to my future research in Louisiana.

As a failed physicist, I'm drawn to the macro and micro view of things—self-similar patterns found on different scales, across the physical landscape as well as through time. The shape of river deltas is found to repeat at scales all the way down to cracks in the mud. We see this in man-made as well as organic structures. A simple underlying mathematics to it all.

In a series of drawings called *Sun Clocks,* I drew the outline of my shadow every hour on the hour. The fundamental, yet often overlooked, action of the Earth spinning on its axis is recorded through the simple action of drawing my own shadow.

*City (New Orleans)* is a small animated drawing that charts the city's growth from its founding in 1718. By scanning the drawing continuously during the process of making it, the finished animation shows the city expanding to its present density and contracting back to its origins in an endless loop, timed to mimic the pace of a human breath—the course of several centuries is condensed to seven seconds. ■

Katie Holten, *Study #96 (Little Carillou Street and 17th Street, Louisiana),* 2012

"THESE AND OTHER ANCIENTLY NATIVE TREES PREVALENT DURING THE EOCENE THERMAL MAXIMUM MIGHT HAVE WHAT IT TAKES TO SURVIVE THE WARMER CONDITIONS WE HAVE ALREADY BEGUN TO EXPERIENCE.

WHY NOT BRING THEM BACK ON A LARGER SCALE TO THE AREAS THEY ONCE LIVED IN, SO THEY CAN FILL THE NICHES THAT WILL BE LEFT BEHIND AS THE MORE HEAT SENSITIVE PRESENT DAY SPECIES BEGIN TO DECLINE?

WOULDN'T THIS BE SOMEHOW ARTIFICIAL? OF COURSE IT WOULD. "
— OLIVER KELLHAMMER

image: *Neo-Eocene forest*, Cortes Island British Columbia, courtesy Oliver Kellhammer

David Gersten

## 32. ARTS, LETTERS & NUMBERS:
### SITUATING ENGAGEMENTS WITH MATERIAL AND EXPERIENTIAL GEOGRAPHIES

all images courtesy of "Arts, Letters & Numbers"

The ubiquitous observation of our time is transformation: cultural, technological, social, political, economical, and geological. We are in the midst of unprecedented re-alignments and re-articulations of every aspect of our lives. There are people and institutions across all disciplines and across the globe that are increasingly confronted by the need for new models of asking the extraordinarily complex questions of our time. The challenges, possibilities of our moment are extraordinary; they call for creative urgency, considered stewardship, and new spaces bringing together diverse voices.

*Arts Letters & Numbers*, a workshop conducted in summer 2012, involving 30 participants from 12 countries, is just such a space. It offered a new approach to constructing alternative pathways of interaction among a wide range of individuals, disciplines and geographies.

Education, by definition, is a transformative pursuit; individuals come together and engage in transformative interactions and experiences. Today, education in the broadest sense holds the capacity of developing new pathways of interaction and forms of knowledge that address the challenges of our increasingly complex world. Each discipline affords us distinct modes of thinking and acting, of articulating: light, substance, space, energy, voice, and thought; they provide elements of perception, comprehension, and engagement. While each discipline presents unique means of comprehending and acting, together they have a shared capacity to, at once, provide the instruments to create transformation and the principles to measure and withstand its consequences. Like many complex structures, such as language or molecular configurations, disciplines are polymorphic. That is to say they can exist in more than one form, they can take on different meanings and organizations depending on their context and environment. In this sense, to gather disciplines within a geographic and intellectual proximity is to create potential for a dynamic disciplinary geography.

One of the core ideas of the Theory of Evolution is this: when individual agents are brought into proximity they interact, building new linkages. Under the right circumstances, these interactions create transformation, developing new forms. Knowledge evolves; creating circumstances of proximity and interaction among a great diversity of agency is fundamental to the emergence of new forms of knowledge. Building new linkages between our spatial, temporal, material, and disciplinary geographies where diverse agencies interact will undoubtedly develop new thought processes and new questions.

The workshop developed as a dynamic series of situations engaging multiple disciplinary, material, and experiential geographies. Taking place outdoors in a large open field, it set up conversations between space, substance, time, light, gravity and memory, people, their work, and the open field. This conversation took place with an acute appreciation of the field itself as a part of

the continuous geography of the solar system. The hope was for individuals to sense the ineffable reality of being within this vast space and time, and to build connections and new linkages between their questions and the spatial temporal geography of the universe. ∎

Notes

1) Circles: Drawing on Friendship started with the perception that we are in the cosmos and the cosmos is in us. Together we asked: How can we create situations that amplify our capacity to listen (to the cosmos and each other) and ask our questions. Working with the elements: fire, water, air and earth, we began to draw them out, drawn them up, this led to a series of comprehensions and actions and then acting and acts. Working with the elements of film, music, voice and bodies we drew on friendship and discovered an emergent libretto between us. The precisions of our emotive embodied knowledge crafted this libretto into a live opera.

2) I have barely caught my breath...40 people from 12 countries spend three weeks renovating an abandoned boarded up building into a place that could work for this project. The construction built an energy and intensity that rolled right into the workshop: Expansive and very close, deep: Each moment was the start of a new story; each photo was the first frame of a new film (lived and shot) sun, fire and filament at the same time. Enormous fires (30 ft in diameter!) then the next day, a master calligrapher from Iran made an enormous Noon with our bodies...then talks on surfing, rap, photography, film, finance, ethics, poetics, race cars, prison, print making, dancers...it just never stopped and it all fed into the work. Deeply human and connected to the cosmos...the big drawing you see hanging in the space is a sun made by the earth and the rain: a huge pile of dirt put on paper outside in the field...then the rain came, the pile of dirt was tall enough that the water did not get to the paper in the middle but only at the edges...so we get an inverse of the earth....a sun made by the earth and the rain. We had at least a thousand such moments...fire and its drawings, found gravel inside...then walking through the gates made by fire...to find tea and touch. Masks from clay cut out of the earth with pick axes, sledge hammers becoming violins, bathrooms becoming cameras obscures, screams caught in glass jars and released later as song, pianos being plucked by a frenzy of hands as they rose in the elevator, water being transferred between bowls by soaking towels and wringing them out; wringing out the tears. All of these Circles were Drawing on Friendship, drawing out human emotions; capturing expressions of our emotive experiences, some in words, but many before and beyond language...a new kind of libretto...a new opera...a new film...a new school.

Oliver Kelhammer

## 33. NEO-EOCENE

Fossil Metasequioa from British Columbia, image by Oliver Kelhammer

> Soon it would be too hot. Looking out from the hotel balcony shortly after eight o'clock, Kerans watched the sun rise behind the dense groves of giant gymnosperms crowding through the roofs of the abandoned department stores four hundred yards away on the east side of the lagoon. Even through the massive olive-green fronds the relentless power of the sun was plainly tangible.
>
> —J.G Ballard, *The Drowned World* (1962)

Back in the mid-1960s, I was a feral sort of a child who loved scampering around construction sites, climbing the huge, grey piles of excavated shale that were popping up all over my rapidly developing Toronto suburb.

I might have been six or so when I first really noticed the slabs of muddy smelling rock often contained the imprints of scallop shells, snails, and fragments of coral, things I recognized from picture books but hadn't yet seen in real life as we lived hundreds of miles from the nearest ocean. Yet 450 million years before, during the Ordovician era, where I was playing would have been the middle of a vast ocean whose limpid, tropical waters teemed with fantastic life forms such as giant, predatory sea scorpions and nautiloids that jetted through the primeval currents like living missiles. I knew this from visiting the Royal Ontario Museum's brand new, McLuhan-inspired, Hall of Invertebrate Paleontology, which recreated detailed dioramas of life in Ontario's ancient seas, complete with theatrical lighting and interactive, taped-looped narratives played through banks of telephone receivers.

So where did it all go, the ocean and its creatures? My world was now just an uninspiring vista of torn-up farmland and under-construction mini malls with the cold, stinking expanse of Lake Ontario looming in the distance, a far cry from the warm coral sea that had left its mark so abundantly all over the local stone. That was when I first began to comprehend that all the things around me I took for granted, the climate I was used to, the location of oceans, even the ground beneath my feet, were completely ephemeral and that someday they would all be just another layer in the palimpsest of time we call geology.

I became more and more fascinated with this idea of planetary change through cataclysm, extinction, and evolution, imagining I could travel back in time to visit those sea scorpions and dinosaurs of bygone geological eras and to experience the world beyond our own species' comparatively recent dominance of the planet. Through every asteroid impact, Ice Age, and super volcanic eruption, biology always seemed eventually to adapt, despite the horrible setbacks, epitomized for me and others of my generation by the extinction of the dinosaurs sequence in Disney's "Fantasia." Yet I knew the demise of the dinosaurs in the late Mesozoic was followed by the rise of mammals in the Cenozoic and so on. But what, I wondered, would come after our own geological age? Looking into the past was one thing, there were rocks and fossils to serve as clues for what had happened, but what about the future?

Lately, this has become more than an abstract discussion. Human-initiated climate change is proceeding so quickly we are now on the threshold of experiencing a planet where conditions are profoundly different from the ones we have come to know and trust. The most agreed upon climate models predict an upward global temperature shift of as much as four degrees Celsius within the next 100 years, with large parts of North America and most of Europe slated to heat up even more than the average. Yet what will this mean? Well we can be certain that organisms and cultures that have evolved within the narrow climatic conditions of recent history will either have to adapt, migrate, or die. The landscape itself is sure to change with the increasing temperature regime that will bring with it drought in some areas and excess precipitation in others. Trees adapted to cool environments will wither in the hotter conditions and pests no longer kept in check by winter frosts will run amok as has already happened in the beetle ravaged pine forests of western North America. It might be already too late to halt this climatological juggernaut, even if our species could express the collective will to do so, which we seem very far from doing. On the contrary, our greenhouse gas emissions continue to rise, pushing potential temperature outcomes still higher.

Metasequoia branch, image by Oliver Kellhammer

Yet the planet has endured such warming before, most recently during the Eocene Thermal Maximum, some 55 million years ago, during which time palms and alligators flourished as far north as Alaska and the Canadian Arctic and the boreal forest was warm and temperate

*Time Landscape, image by Oliver Kellhammer*

and far more biologically diverse than it is now. Then a long term cooling trend started and many of the tree species that had been widespread gradually disappeared from most of their range, surviving only in isolated refuges in areas that are now the southeastern United States and Western China, where conditions stayed just stable enough for them to hang on. Some formerly common trees such as *Metasequoia* and *Gingko* were almost completely wiped out, reduced to a few isolated groves of a few thousand trees, despite having once dominated large swaths of the northern hemisphere's forests. Known previously only from fossils, the *Metasequoia* was only discovered as a living species as recently as 1943, when botanists found a small population of them surviving in China's Hubei province. They collected seeds and now the tree is widely cultivated. Similarly, the *Gingko*, whose petrified remains can be found as far afield as Scotland and Oregon, vanished from all its native haunts except for another tiny area of China from where it was disseminated by Buddhist monks, who revered it and planted it in their temple gardens. Attractive and adaptable, Gingkos are now widely grown in parks and gardens and are popular as a pollution resistant tree in cities.

Though we have done incalculable damage to the planet, we humans at least can congratulate ourselves for snatching the *Ginkgo* and the *Metasequoia* from the brink of extinction and, through horticulture, giving them a new lease on life in our parks and gardens. But why stop there? These and other anciently native trees prevalent during the Eocene Thermal Maximum might have what it takes to survive the warmer conditions we have already begun to experience. Why not bring them back on a larger scale to the areas they once lived in, so they can fill the niches that will be left behind as the more heat-sensitive present-day species begin to decline? Wouldn't this be somehow artificial? Of course it would. Yet modifying our earth's environment has already been our species' greatest legacy—so much so that the age in which we now live has been dubbed the Anthropocene in reference to the ubiquity of our impact. As well as affecting living systems, *Homo sapiens* have become the planet's most powerful geomorphological engine, expending an amount of energy annually in our re-shaping of the earth's crust that is roughly equivalent to the natural processes of erosion and mountain building. On top of the fossilized reefs and bone beds of previous geological ages, we are piling our own strata of abandoned architecture and persistent artificial materials, such as plastic. We leave behind anthromes in place of biomes and even Earth's so-called "natural" areas now exist only under the precarious fiat of our political systems or because we haven't yet gotten around to incorporating them into our ever more voracious supply chains. In view of these epic and overarching realities, the reintroduction of a few formerly native tree species seems an almost minor sort of intervention and to my mind definitely worth a try. As an artist, I was interested in the social nature of such a proposition. Addressing climate change through strategic botanical intervention seemed like a proactive alternative to the hand wringing and negativity that have frequently characterized this issue.

I was first drawn to the aesthetic potential of tree planting as a time-based art practice back in 1979, when, as a young artist visiting New York City, I first encountered Alan Sonfist's newly installed *Time Landscape*. For those not familiar with it, *Time Landscape* is an assemblage of horticultural plantings in a small lot just off Houston Street where Sonfist attempted to replicate a portion of Manhattan's original vegetation from the time before European colonization. When I first saw it, *Time Landscape* was still a rather abject arrangement of newly planted shrubs and saplings but over the intervening 30 years I watched it gradually transform into a stately, sylvan grove and it now looks as if it has always been there, which of course it hasn't. But then again it has, as the horticultural landscape is really just a re-setting by the artist of the state of the place at an earlier moment in time. What's different is the reversed figure/ground relationship between the once dominant biome and the architectural anthrome that has so overwhelmed it. By returning the site to its ecological past, Sonfist made time fluid and one can easily imagine oneself moving back and forth through the layers of its architectural and botanical history. But the re-installed former past has also moved

Ten-year-old Metasequoia, Cortes Island, British Columbia
image by Oliver Kellhammer

forward, which makes *Time Landscape* a kind of anti-monument whose elements have their own agency and change constantly in appearance. As the trees grow on at their own stately paces, adding growth rings and extending their branches, they serve as a kind of time code, independent of our own frenetic aspirations.

But I wanted to turn the clock back more than just a few hundred years. Recreating the forest landscape of the Eocene was an even more conjectural undertaking. Fortunately the plants of the Eocene have left a rich fossil record that is well documented. By the late 1990's, I had established test plantings in my Cortes Island yard of many of the tree genera whose fossils have been found in British Columbia's Eocene deposits. Though I was limited for space, I was able to grow specimens of the conifers: *Metasequoia*, *Sequoiadendron* (Coast Redwood), *Sequoia* (Giant Sequoia), *Cunninghamia*, *Sciadopitys*, and *Ginkgo*, as well as deciduous genera such as *Juglans* (walnut) and *Ulmus* (elm). The trees thrived with minimal care and within a few years I was eager to scale up the project to a landscape level by introducing entire groves of them to an existing forest ecosystem.

The opportunity to do so came about in 2008, when I first met the UK-based botanist Rupert Sheldrake, who had recently purchased an acreage on the other side of our relatively small island. The bulk of his property had been logged-out a few years earlier by a timber company and then partially replanted in Douglas Fir *(Pseudotsuga menziesii)* and Western Red Cedar *(Thuja plicata)*, the most commonly prescribed species for reforestation in the region. There were, however, substantial areas that hadn't yet been planted, thus providing an ideal situation for me to try out larger numbers of some of the Eocene tree varieties I had been experimenting with. With Rupert's financial and scientific sponsorship, the experiment proceeded apace and together with a professional forestry worker I planted hundreds of seedlings in a two year period. Despite unprecedented, back-to-back, summer droughts in 2008 and 2009, many of the trees thrived, most notably the *Sequoiadendron*, *Sequoia* and *Juglans* (*cineria* and *nigra*). Though stressed by drought, a number of specimens of *Metasequoia* and *Gingko* also made it through and are now putting on new growth. The *Metasequoia* is adapted to seasonal flooding and might gain a competitive advantage if the climate were to become wetter as well as warmer.

Unbeknownst to me, just after the property had been clear-cut by the timber company, it had been placed under a well-intentioned covenant held by an organization known as Nature's Trust. Though it allows for extensive industrial logging, the document stipulated that any subsequent reforestation be comprised of species native to the site's presently extant CWHxm1 (Coastal Western Hemlock, very dry) biogeoclimatic zone; essentially a seasonally dry rainforest. By the time we discovered this, the Neo-Eocene plantings were already well underway and Rupert and I proceeded to negotiate with Nature's Trust from our position that in the context of rapid climate change, the term "native" should be expanded to incorporate formerly, even prehistorically, native species, given the distribution of present day biogeoclimatic zones is surely likely to change. Originally, Nature's Trust deemed such trees as *Sequoia* and *Metasequoia* "exotic" and "potentially invasive" and wanted them removed, but in the end we worked out a compromise that allowed these anciently native species to be introduced to no more than 30% of the property's total area.

Predicting the future is always a risky business yet good science and the human imagination can give us at least some idea of what we might be in store for us with an increasingly hotter planet. In my mind, I like to travel through time to maybe a hundred years from now to try and picture how my "Neo-Eocene" plantings might be doing. If the temperatures continued to rise and the summers remained relatively dry, the Cortes Island landscape might more resemble that of present day coastal California. The redwoods and walnuts I planted might then be thriving within an otherwise depauperate forest where only some of the more drought tolerant current species, such as Douglas Fir and Arbutus (*Arbutus menziesii*) would still manage to hang on. Alternately, if it got warmer and wetter, particularly in the summer, the landscape could evolve into something resembling northern Florida, with a more humid regime favoring

Rupert Sheldrake with young Coast Redwood, image by Oliver Kellhammer

*Metasequoia* and *Gingko*, perhaps even palms and cycads. The fact is we just don't know how climate change will turn out, but it makes good sense to hedge our bets. Human beings have a long history as vectors, spreading other species around the planet with both harmful and beneficial outcomes. To my mind we'll have to reassess some of our more cherished notions on what constitutes a native ecology and adopt a more geological perspective as to which species belong where, so we can invite some of Earth's most magnificent trees back to the places they once inhabited when conditions last suited them. Not to do so would be to turn our back on a precious biological heritage. Many tree varieties once more widely distributed during past warm periods now occupy extremely limited and geographically isolated ranges. Human-assisted migration could allow them once more to be a part of a broader ecological framework. With the rise of global warming and the Anthropocene, the future of Earth's biodiversity is already in our hands. Rightly or wrongly, we have taken on strangely god-like powers to affect the world's climate and the evolution of its ecosystems. ■

Lisa Hirmer

## 34. THE LESLIE STREET SPIT

all images, Lisa Hirmer, *Untitled*, from *Leslie Street Spit Geology*, 2012

The Leslie Street Spit is a long finger of artificial land stretching into Lake Ontario from Toronto's shore. It was originally conceived in the 1950s as a breakwater for the city's northern harbor, but with changes in shipping methods quickly became obsolete in its original purpose. It remained, however, such a convenient place to discard of what is cleverly called "clean fill"—referring mostly to broken up pieces of demolished urban buildings—that it continues as an active dumping site today, expanding ever further into the lake.

Owing to its open land and remote location, the spit was rapidly colonized by vegetation and wildlife, including several endangered species. Though this was never part of official plans, the spit became a hybrid-urban wilderness where, amidst the rubble of brick and bent rebar, nature thrived.

These photographs document one of the oldest sections of the spit where the dumped rubble has settled and compacted into a soil like state. On the surface, where vegetation abounds, the ground reads as natural, not much different from any other conservation area. But, along the shoreline where the water pulls at the edge of the spit, the contents of the ground become visible—a rust colored metallic band, speckled with shards of glass, plastic and other debris. Essentially made from old iterations of the city, disposed of to make way for the new, this band offers an early glimpse of the urban world made geologic. ■

# SECTION 4: GEOLOGIC TOMORROW: WILD AND POTENT FUTURES

Shimpei Takeda

## 35: TRACE — CAMERALESS RECORDS OF RADIOACTIVE CONTAMINATION

As an artist working with analog photographic techniques that are rapidly in decline, a native of Japan, and moreover being born in Fukushima prefecture, I took on a mission to create a physical record of the worst man-made nuclear accident in history.

A photograph can be created by electromagnetic radiation other than light. Radiation and visible light are similar forms of energy, they just travel in different wavelengths. Using cameraless processes, I want to capture the current state of Japan directly, by exposing photosensitive material to traces of radiation emitted from contaminated particles.

Through visualizing the traces into visible form, the resulting images will speak to us beyond the photograph, and perhaps they will be a valuable asset and documentation for future generations.

### EXPOSURE PROCESS

Radiation such as X-rays and gamma rays are high-energy electro-magnetic waves that can expose light-sensitive film and paper. Shown below, radioactive substances are placed on top of the film, emitting radiation that passes through a plastic sleeve to create a trace.

## PRELUDE (PRE-EXPERIMENTATION) | JULY - AUGUST, 2011

Before collecting soil in Japan, I needed to test a few things. Making a radiograph, exposing photosensitive paper with radioactive materials, is more common than you might think. I've found many enthusiasts doing related experiments. Normally, taking an x-ray photograph requires an x-ray generator emitting high radiation to expose x-ray film in a very short period of time in order to get a clear image and to minimize the radiation exposure. But, if you use instant film, you can make a radiogram without having a darkroom, developing trays, or chemicals.

A Photogram, placing an object on photo paper to expose the shadow of shapes, requires from 1/10000 to a few seconds of exposure. And following that procedure, processing the paper takes about five minutes. With this relatively short time process, although it's hard to get the exactly same composition each time, you can create similar compositions with adjusted exposure time or brightness of the light source. In contrast, creating autoradiograph by using low level contaminated soil takes a very long process. In my calculations, and from my experimentation, I project that, by using the soils that I am planning to use, the exposure required will be a few weeks to a few months to get decent dark gray. And if you need to redo it to get longer exposure, you need to redo from zero again.

## EXPERIMENT 1: RADIOACTIVE ISOTOPE DISK (Cs-137)

Through my research on radiograms, I found that I can purchase an exempt quantity of nuclear isotopes in a safely sealed container, for scientific or educational research purposes. I chose a Cesium-137 disk, one of the main isotopes found in nuclear fallout. The disk was "freshly" shipped from a nuclear supplier in Tennessee. It is sealed in one-inch diameter plastic with epoxy to prevent leakage and contamination. I placed the disk on four layered pieces of four by five inch enlarging papers and kept them in a blackout box. A week later, I developed the papers, and the Cesium's activity was clearly recorded. As a sheet of paper slightly blocked the

From *Trace – camerahess records of radioactive contamination*, Shimpei Takeda, 2012

radiation, the lower sheets got a smaller black dot. The Geiger counter that I use for now is not the most sophisticated one, but it reflected the radiations at about 6.7μSv/h.

*At a later date, using a more sophisticated radiation meter, I put the disk in front of Mica window, right outside of the Geiger tube and I found the disk's reading was actually over 200μSv/h.

## EXPERIMENT 2: CRUSHED RADIOACTIVE RED FIESTAWARE (U-235 & U-238)

Fiestaware was a popular vivid colored dinnerware introduced in 1936. It wasn't the first solid color tableware company, but it was the first widely mass-promoted and marketed in the U.S. Fiesta's red-orange glaze (1936-43) is well-known for containing significant amounts of radioactive materials in its uranium oxide glaze. Actually, it was used widely for the red glaze produced by all U.S. potteries of the era. The use of uranium in its natural oxide form dates back to at least the year 79 CE, when it was used to add a yellow color to ceramic glazes. Yellow glass with 1% uranium oxide was found in a Roman villa on Cape Posillipo in the Bay of Naples, Italy. When 1% Uranium oxide is added into the glaze, it becomes yellow, and the more added the more orange to dark red it becomes.

This radioactive Fiestaware that I purchased is believed to contain 15% uranium oxide in the glaze, which is 0.17oz (5g) per plate in average. In 1943, the government banned using Uranium by corporations, and confiscated the Uranium of the Homer Laughlin China Co. It reminds me of Imperial Japan's National Mobilization Law that collected all metals from citizens, including bells in Buddhist temples for war efforts. Surely, the scale is much different, however. In any case, radioactive Fiestaware has been described as one of the most radioactive commercial products you could buy.

For this experiment, prior to exposing photo-sensitive materials with nuclear contaminated soil, I crushed the Fiestaware plate to make the material more random than a single flat plate. I stored an enlarging paper with the crushed plate in a box and exposed it for a week. The exposure result was the shapes of the plate pieces outlined on the paper. I determined an even longer exposure is needed to get dark grays. As I used my Geiger counter to measure this material, it indicated 1.47µSv. After a month in the enclosure, the Uranium Oxide's radiation is clearly recorded on the photo paper.

## CHOOSING SOIL COLLECTING LOCATIONS | AUGUST - SEPTEMBER, 2011

Shimpei Takeda at Nakano Fudoson in Fukushima-city, Fukushima. Photo by Keisuke Hiei, 2012.

Fortunately, my family, close friends, and their families were not directly injured by the devastating earthquake and tsunami, which hit Japan on March 11th, 2011. While the world and media watched the country as it was devastated by the disaster, many of my friends kindly called or e-mailed me to check if my family and friends were affected. As it's been almost a year and a half since last March, the media has been busy covering the economic fallout; while the physical radiation fallout, and potential for more disaster continues. I recently saw a photo of the aftermath, taken by a good friend of mine from Sendai. Some damaged neighborhoods haven't recovered at all. Massive debris has barely been cleaned up and radiation is still being emitted. In times like this, I feel I'm alive because my ancestors survived though numerous disasters, wars, and starvation. From this perspective, when I look at a map and at historical sights, it's hard not to imagine the historical blood marking the ground. Tombs, graveyards, a temple, or a battlefield are direct marks, but also some war-related ruins and other hidden memories of tragedies can often be found everywhere in Japan.

I am not expecting that a radioactive trace taken at these historical sights will serve as a memorial service. More, it would be a meaningful pilgrimage to visit these sights and conduct the trace upon each visit.

## SOIL COLLECTING TRIP | JANUARY, 2012

In January 2012, accompanied by Shingo Annen, a hip-hop activist, and architect Keisuke Hiei, the "Trace" crew collected 16 soil samples in 12 locations. I selected distinctly different locations in five different prefectures, all of which contain a strong memory of life and death, such as temples, shrines, war sites, and ruins of castles. I included Sukagawa City, forty miles away from the Fukushima nuclear plant, my birthplace and current home to my grandparents.

Radiation was measured with Radalert 100, a handheld Geiger counter, at approximately four feet from the ground for air measurement, and directly on the ground for ground-level measurement.

## SOIL SAMPLE DATA

Kegon Falls (Nikko, Tochigi)
12/31/2011, Sunny. Air: 0.249μSv/h /
Ground: 0.446μSv/h

Former Kashiwa Military Airbase (Kashiwa, Chiba)
1/2/2012, Sunny. Air: 0.519μSv/h /
Ground: 0.623μSv/h

Former Kasumigaura Naval Air Force (Ami, Ibaraki)
1/3/2012, Sunny. Air: 0.415μSv/h /
Ground: 1.007μSv/h

Shioyasaki Lighthouse (Iwaki, Fukushima)
1/3/2012, Sunny. Air: 0.228μSv/h /
Ground: 1.152μSv/h

Chūson-ji (Hiraizumi, Iwate)
1/7/2012, Snow
Air: 0.321 / Ground: 0.45μSv/h

Hyaku-Shaku Kannon (Soma, Fukushima)
1/8/2012, Sunny
Air: 0.633 / Ground: 2.637µSv/h

Lake Hayama (Iitate, Fukushima)
1/8/2012, Sunny
Air:1.848 / Ground: 6.438µSv/h

Iwase General Hospital (Sukagawa,
Fukushima) 1/5/2012, Sunny. Air: 0.363µSv/h /
Ground: 0.560µSv/h

Nakano Fudoson, buddhist temple
(Fukushima, Fukushima) 1/5/2012, Snow.
Air: 0.67µSv/h / Ground: 1.030µSv/h

Nihonmatsu Castle (Nihonmatsu, Fukushima)
1/4/2012, Sunny. Air: 1.910µSv/h /
Ground: 4.299µSv/h

Asaka Kuni-tsuko Shrine (Koriyama,
Fukushima) 1/4/2012, Sunny. Air: 1.152µSv/h /
Ground: 3.780µSv/h

Kamayama Limestone Quarry
(Tamura, Fukushima) 1/5/2012,
Sunny. Air: 0.280µSv/h /
Ground: 0.415µSv/h

## SELECTED LOCATION'S INFO

### 1. Kegon Falls
Standing at 97m (318ft) high, Kegon Falls is one of Japan's three highest waterfalls. In 1903, an elite student committed suicide into the fall, which has since seen 185 followers to attempt the same (40 succeeded). Ever since, it is known as a suicide spot.

### 3. Former Kasumigaura Naval Air Force
In 1922, Kasumigaura naval air force was opened, and it focused on flight education for pilots. The site is currently used as Japan Ground Self-Defense Force. On the premises, dozens of tanks used in WWII are on display. There is also a memorial hall exhibiting young soldiers' portraits, notes, and belongings.

### 4. Shioyasaki Lighthouse
Shioyasaki Cape was known as the hardest area to pass in the region, and many ships were sunk around the cape. It's been said that in the 1850's a fire beacon was setup as a mark for sailing. Shioyasaki Lighthouse first opened in 1899. During WWII the lighthouse was heavily damaged by gunfire and became derelict. On August 10th, 1945, a 21-year-old operator manning the lighthouse was killed under machine-gun fire.

### 5. Nihonmatsu Castle
Built in 1414, during the Boshin War, a civil war battle was also fought at Nihonmatsu Castle, from 1868 to 1869, between forces of the ruling Tokugawa Shogunate and others seeking to regain to political power in the imperial court. The Castle's reputation became as a place of unsuccessful defense. Failure pervaded Nihonmatsu, and teenage youth corps were forced to fight there.

### 6. Asaka Kuni-tsuko Shrine
In 135 AD, this shrine was built along with the new province, and meant to be dedicated to God of farming of in Japanese mythology. Back then, a province official was also in charge of spiritual and religious affairs, specifically the Shinto rites of each province. In later ages, Commander-in-Chief of the imperial court prayed for victories against the native population, pushing them further north.

### 7. Iwase General Hospital
Opened in 1872, Iwase General Hospital was one of the earliest hospitals to introduce western medicine outside of Tokyo at the time. There are some records showing that statesmen in the Meiji era came there for a round of inspection. Over 100 years later, I was born at this hospital.

## NOTE | FEBRUARY, 2012

Soil Collection & Setup for exposure (Jan – Feb, 2012)

While the "Trace" crew collected 16 soil samples in 12 locations, we carried radiation meters with us during the entire trip. As we traveled to my birthplace in Fukushima, my parents' house in suburban Tokyo, where I grew up, and to all the memorial places we visited, I found these places all to be contaminated, albeit in different degrees. Although I was aware of this previously from data visualization maps, looking at the raised numbers on my Geiger counter as I traveled to these locations confirmed this to me, sadly.

Since the Fukushima Daiichi nuclear disaster occurred, I've studied the effects of radiation on humans intensively. Hiroshima and Nagasaki's nuclear bombs, numerous nuclear weapons testing in the atmosphere and underwater, and the Chernobyl disaster—these events created enough guinea pigs to fill the reports on the effects. Visiting contaminated areas and seeing the number rising on radiation meters, it sometimes made my heart beat fast. But, ironically, once you ignore the numbers, you just become aware of the beautiful country's landscape spread in front of you.

This is why I wanted to visualize this nuclear disaster in my artistic processes. Sixteen soil samples have been stored with 8×10 B/W sheet films in a light-tight enclosure individually for a month, and I developed the film to find out how the Fukushima disaster has created the Trace. ■

From *Trace — cameraless records of radioactive contamination*, Shimpei Takeda, 2012

Jamie Kruse

## 36. THE POWER OF CONFIGURATION: WHEN INFRASTRUCTURE GOES OFF THE RAILS[1]

> . . . things that never happened before are possible. Indeed, they happen all the time.
>
> —Charles Parrow, Bulletin of Atomic Scientists

Yokota Airbase, Japan: Airmen scan members of the Queensland Urban Search and Rescue Taskforce for Radiation, March 20, 2011. Image: courtesy of U.S. Air Force Photo: Master Sgt. Kimberly Spinner

On November 11, 2011, the Institute of Nuclear Power Operations (INPO) delivered a timeline to an audience of U.S. industry executives, the Nuclear Regulatory Commission, and members of Congress. It detailed the unfolding of events at the Fukushima Daiichi nuclear power station during, and in the critical hours following, the earthquake and tsunami of March 11, 2011. The story contained within the document entitled, "Special Report on the Nuclear Accident at the Fukushima Daiichi Nuclear Power Station," reads like a screenplay. The gravity of the immediate hours after the accident, detailed and delivered through stark legalistic language, confronts the reader with a poignant reminder: when things don't go as planned, human bodies are what must show up to try to steer things back on course.

The INPO report serves well as a reference manual for contemporary designers, architects, urban planners, and engineers. It's a powerful invitation to consider *configuration* in relation to infrastructure design and planning. Configuration: how and where we stage what we design, how humans and infrastructures are situated in relation to one another, and how they inevitably interact. Yet humans and infrastructures exist not only in relation to one another and the landscape, but also in relation to the multitude of earth forces capable of rising up and challenging our best design and engineering capacities.

TEPCO, the electric company that owned the Fukushima Daiichi plant, clearly had never imagined, nor planned for, the spectacular forces that materialized on March 11, 2011. Tsuneo Futami, a nuclear engineer who was the director of Fukushima Daiichi in the late

---

1  A version of this essay appeared on the *Friends of the Pleistocene* blog (fopnews. wordpress.com) in January 2012. A version was also presented at the public symposium entitled *Energy!* hosted by the Vera List Center for Art and Politics in March 2012 in conjunction with the installation, *Thingness of Energy* (Jamie Kruse) at the Sheila Johnson Designer Center, Parsons The New School for Design, February-March 2012.

1990s, has stated, "We can only work on precedent, and there was no precedent. When I headed the plant, the thought of a tsunami never crossed my mind."

In the crucial minutes and seconds in which the unimaginable actually does unfold, highly complex situations transpire in instants. The configuration of people and things becomes fundamental to what happens. Reality shifts into a configurationist map-in-motion, a rolling landscape of trip wires and tipping dominoes. Each action, placement and location sends vectors of consequences propagating into the far future.

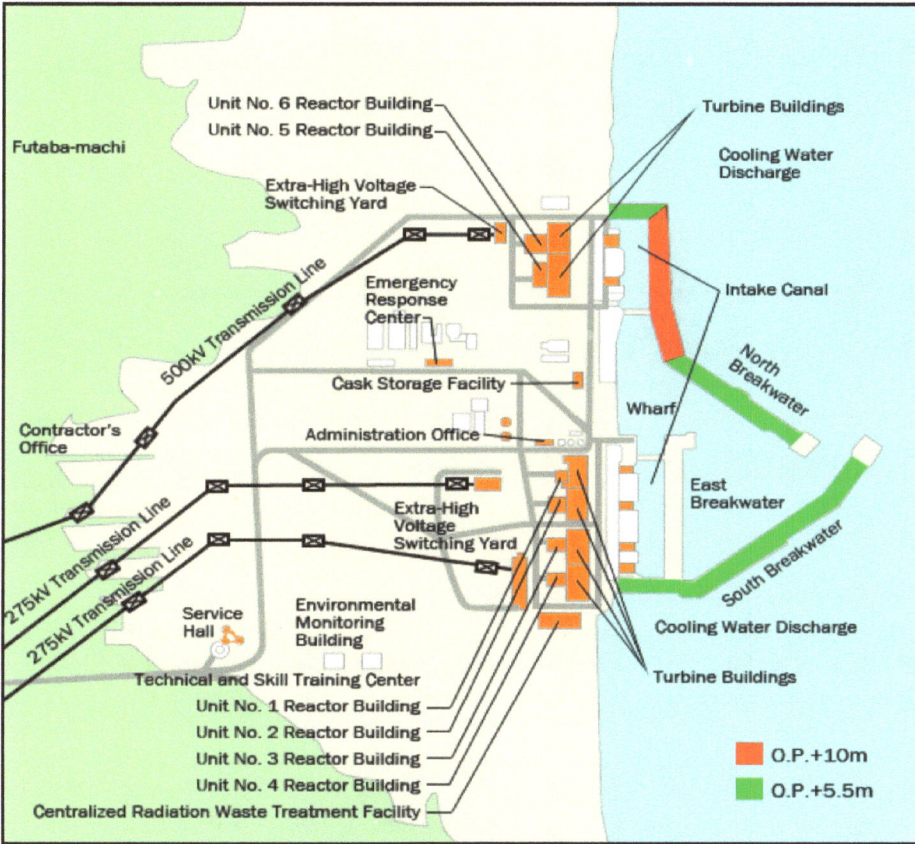

"Configuration" diagram of the Fukushima Daiichi facility
Image courtesy of the Institute of Nuclear Power Operations.

How high do the waves reach up the bluff? Where are the fire trucks parked in town? How far inland do the dry casks sit? How deep do the spent fuel rods rest in their cooling pools? How far are the plant's exit gates from here?

Where each building is sited in relation to each human, each fire truck, each monolithic sea wall, each wave—all matter immensely *and* in interconnected ways.

This highly particular configuration of things varies from moment to moment and is what will mix and ramify into the resulting "nexts." Water is no longer water, but a substance flooding basements storing backup generators. Utility employees become authorities, emergency personal, evacuees, heroes, and victims.

At Fukushima, the force and scale of the events reconfigured an energy generating plant, typically a life-supporting affordance, into something else, into something unrecognizable. Fukushima Daiichi shifted from being an energy-producing infrastructure into a risk-generating machine—with massive geologic consequence. The facility dissolved into an assemblage of unpredictable actants in the forms of zirconium rods, uranium, plutonium, hydrogen, salt water, copper, plate tectonics, massive waves, electricity, darkness, design ingenuity, engineering failure, steel, reinforced concrete, basements, and sea walls.

These elements not only performed independently of human desire, they acted back upon us in profound and irrevocable ways. Many of these things-become-forces continue to do so today, more than a year later. Some will continue to do so for generations to come.

Japan, March 25, 2011. Image: courtesy of U.S. Air Force. Photo: Tech. Sgt. DeNoris A. Mickle.

These actants compose the shifting trajectories of global energy futures as they birth new actants such as long-term power outages, evacuations, containment failures, explosions, aftershocks, media coverage, government regulations, industry investments, energy dependency, changes in public opinion.

An awareness is growing: long-term futures hinge on how we humans assemble with things that can unexpectedly reconfigure in a matter of micro-seconds. We are beginning to sense the necessity of acknowledging forces that extend deep below the earth's surface, and the relevance of timescales that exceed human time. As we begin to respond, we are prompted to consider: *what if anticipating geologic scales of force, change and effect became a common design specification? What if energy production, policy-making, and infrastructure design projects began to account for lively and wildly unpredictable geologic actants?*

Excerpts from the "Unit 1 Validated Event Timeline" (pages 71-80) in "Special Report on the Nuclear Accident at the Fukushima Daiichi Nuclear Power Station." All times are provided in Japan Standard Time (JST).

**11-March 14:46**
Automatic reactor scram signal on seismic trip

**11-March 14:47**
Automatic turbine trip on high vibration

**11-March 14:47**
6.9-kV bus 1D power loss

**11-March 14:47**
6.9-kV bus 1C power loss

**11-March 15:27**
The first wave of a series of tsunamis, generated by the earthquake, arrived at the station.

**11-March 15:35**
The second tsunami hit the station.

**11-March 15:37**
Loss of all AC power occurs. Instrumentation and emergency systems gradually fail between 1537 and 1550.

**11-March 15:37**
The loss of DC distribution systems results in the loss of control room indications and alarms.

**11-March 15:37**
The control room lighting was lost and only emergency lighting remained.

**11-March 15:3**
The control panel indications for HPCI were barely lit but slowly faded to black.

**11-March 15:42**
TEPCO entered its emergency plan because of the loss of all AC power, in accordance with Article 10, paragraph 1 of the Nuclear Disaster Law. Government offices were notified. The corporate Emergency Response Center was established.

**11-March 16:36**
Temporary batteries and cables were gathered and carried to the units 1 and 2 control room. After confirming the wiring layout using drawings, batteries were connected to instrument panels.

**11-March 20:07**
Because there were no working indications in the control room, operators checked reactor pressure locally in the reactor building. Reactor pressure was 1,000 psi (6.9 MPa).

**11-March 20:50**
Authorities of the Fukushima prefecture ordered evacuation of the population within a 1.2 mile (2 km) radius of Fukushima Daiichi.

**11-March 21:19**
Water level indication was restored in the control room. Reactor water level was approximately 8 inches (200 mm) above the top of active fuel (TAF).

**11-March 21:51**
Access to the reactor building was restricted because of high dose rates.

**11-March 23:00**
A radiation survey identified dose rates of 120 mrem/hr (1,200 μSv/hr) in front of the reactor building north door on the first floor of the turbine building and 50 mrem/hr (500 μSv/hr) in front of the door to the south. The government authorities were notified at 2340.

**12-March 00:06**
In the control room, operators assembled piping and instrumentation drawings, the accident management procedures, valve drawings, and a white board. The operators began to develop a procedure for venting, including how to manually operate the valves, and the associated sequence.

**12-March 01:48**
At some point, the installed diesel-driven fire pump that was standing by to pump water into the reactor malfunctioned. In an attempt to restart the fire pump, diesel fuel was carried to the pump and the fuel tank was refilled, and batteries stored in an office were carried to the room and installed; but the pump would not start. Workers began considering using fire trucks to supply water to the plant fire protection system.

**12-March 01:48**
The station had three fire engines, but only one was available to support injecting water into the Unit 1 RPV. One fire engine was damaged by the tsunami. The second fire engine was at parked adjacent to units 5 and 6 but could not be driven to Unit 1 because earthquake damage to the road and debris from the tsunami had restricted access between units 1 through 4 and units 5 and 6.

**12-March 02:24**
In preparation for manually venting the containment, a radiological evaluation of working conditions in the torus room was provided to the ERC. With radiation levels at 30 rem/hr (300 mSv/hr), workers were limited 17 minutes of time in order to remain below the emergency response radiation limit of 10 rem (0.1 Sv). Workers were required to wear a self-contained breathing apparatus (SCBA) with a 20-minute air supply and would be given potassium iodide tablets.

**12-March 03:45**
Workers attempted to enter the reactor building airlock door to perform surveys. As soon as the door was opened, workers saw steam and closed the door. No surveys were performed.

**12-March 04:30**
The ERC informed the control room that field operations were prohibited because of tsunami warnings.

**12-March 04:50**
Workers were directed to wear full-face masks with charcoal filters and coveralls when in the field.

**12-March 05:44**
Radiation levels at the site boundary increased, and the Prime Minister expanded the evacuation zone around Fukushima Daiichi to 6.2 miles (10 km).

**12-March 05:52**
A total of 264 gallons (1,000 liters) of fresh water was injected via the fire protection system.

**12-March 06:30**
A total of 528 gallons (2,000 liters) of fresh water was injected via the fire protection system.

**12-March 06:33**
TEPCO confirmed that some residents of Ookuma-machi, which is inside the evacuation zone, had not evacuated yet. The residents had not left because they were not sure in which direction to evacuate.

**12-March 07:11**
The Prime Minister arrived at the station.

**12-March 08:37**
The Fukushima Prefectural government was informed that venting would start at approximately 09:00. Venting was being coordinated to ensure the evacuation was completed prior to venting commencing.

**12-March 09:03**
The control room operators formed three teams to perform the venting, with two operators on each team (one to perform actions and the other to assist by holding flashlights and monitoring dose rates and for other safety concerns, such as ongoing aftershocks). Because there was no means of communicating with the field teams, the decision was made to dispatch one team at a time, with the next team leaving only after the preceding team returned.

**12-March 09:30**
The second team of operators was unsuccessful in the attempt to manually open the suppression chamber air-operated vent valve. The operators entered the torus room but had to turn back because they expected they would exceed their 10 rem (100 mSv) dose limit.

**12-March 11:39**
The government was notified that one of the operators who had entered the torus room to attempt to vent the PCV had received 10.6 rem (106 mSv) radiation dose.

**12-March 15:36**
A hydrogen explosion occurred in the reactor building (secondary containment).

**12-March 15:36**
The explosion caused extensive damage to the reactor building and injured five workers. Debris ejected by the explosion damaged the temporary power cables, along with one of the large portable generators. The temporary power supply for the standby liquid control system and the hoses that had been staged for seawater injection were damaged beyond use. Although the fire engines were damaged, they were still usable. The injured workers were carried to safety. Station workers, including the personnel working on the standby liquid control system and laying temporary power cables, had to evacuate for an accountability. The area surrounding Unit 1 was strewn with highly radioactive debris, so cleanup required support from radiation protection personnel.

**12-March 16:27**
Radiation dose rates at the monitoring post reached 101.5 mrem/hr (1,015 μSv/hr), which exceeded the 50 mrem/hr (500 μSv/hr) limit specified in Article 15, clause 1 of the Act on Special Measures Concerning Nuclear Emergency Preparedness (abnormal increase in radiation dose at the site boundary). This was reported to the authorities.

**12-March 18:25**
The Prime Minister ordered the population within 12.4 mile (20 km) radius of Fukushima Daiichi Nuclear Power Station to evacuate.

**12-March 18:30**
Field inspections revealed the area around the units was littered with debris, and the equipment that had been staged to provide power to the standby liquid control system and hoses staged to inject seawater had been damaged and were no longer usable.

**12-March 19:04**
The injection of non-borated seawater into the reactor commenced using the fire engines.

**24-March 11:30**
Lighting was restored to the units 1-2 control room.

**25-March 15:37**
Reactor injection was changed from seawater to fresh water.

Forty-one minutes after the earthquake, at 15:27, the first of a series of seven tsunamis arrived at the site. The maximum tsunami height impacting the site was estimated to be 46 to 49 feet (14 to 15 meters). This exceeded the design basis tsunami height of 18.7 feet (5.7 meters) and was above the site grade levels of 32.8 feet (10 meters) at units 1-4 (p. 3).

The tsunami design basis for Fukushima Daiichi considered only the inundation and static water pressures, and not the impact force of the wave or the impact of debris associated with the wave. The design included a breakwater, which ranged in height from 18 ft (5.5 m) to as high as 32.8 ft (10 m), as shown in Section 1.1 (p. 47).

The Act on Special Measures Concerning Nuclear Emergency Preparedness (commonly referred to as the Nuclear Disaster Law) was established in 1999 in response to the September 30, 1999 inadvertent criticality accident at the Tokai uranium processing plant. The accident resulted in overexposure of three plant workers and additional unplanned exposures to 66 plant workers, local inhabitants, and emergency support personnel (p. 67). ■

---

References

Bennett, Jane. 2011. *Vibrant Matter: A Political Ecology of Things*. Durham: Duke University Press.

Onishi, Norimitsu and James Glanz. 2011. "Japanese Rules for Nuclear Plants Relied on Old Science," *The New York Times*, 26 March, http://www.nytimes.com/2011/ 03/27/world/asia/27nuke.html.

Perrow, Charles. 2011. "Fukushima, Risk, and Probability: Expect the Unexpected," *Bulletin of the Atomic Scientists*, 1 April, http://www.thebulletin.org/web-edition/features/fukushima-risk-and-probability-expect-the-unexpected.

Institute of Nuclear Power Operations. 2011. "Special Report on the Nuclear Accident at the Fukushima Daiichi Nuclear Power Station," Nuclear Energy Institute, Accessed 11 November, http://www.nei.org/resourcesandstats/documentlibrary/ safetyandsecurity/reports/special-report-on-the-nuclear-accident-at-the-fukushima-daiichi-nuclear-power-station.

"The design we must do in an age of hyperobjects will inevitably take them into account, because we can't unthink our knowledge of them. This means that design must account for thousand, ten thousand, and hundred thousand year timescales. It must account for Plutonium 239, which remains dangerously radioactive for **24,100 years.**"

—Tim Morton, "Zero Landscapes in the Time of Hyperobjects"

TRUCK TRANSPORTING TRANSURANIC WASTE, I-15, IDAHO
image: smudge studio, 2010

Bryan M. Wilson

## 37. THE NUCLEAR PRESENT

On September 29th, 2009, I was about to begin a pilgrimage to a remote location in the Southwestern desert. With my mother as navigator, my plan was to drive from my home state of Montana to the Trinity Nuclear Test Site, located within the White Sands Missile Range between Alamogordo and Soccorro, New Mexico. The United States military opens the Trinity Test Site to the general public two times a year, in the beginning of October and April. On July 16, 1945, the first plutonium-based atomic weapon was detonated on a remote location in the New Mexican desert. The subsequent explosion of "The Gadget" was described as a blinding light of golden, purple, violet, gray, and blue, lighting up the early desert morning as if it were high noon. At ten miles away, spectators required dark sunglasses to look upon the explosion without damage to their eyes.

The Trinity explosion did not produce a crater at ground zero, but left behind a small lake of greenish glass, dubbed "trinitite." It is theorized that the intense heat generated by the nuclear fireball (in excess of 14,000 degrees Fahrenheit) liquefied the desert sand floor into glass and sucked the material into the emergent mushroom cloud. At such high force and temperature, the vaporized glass behaves like water in a typical cloud: it collects in the mushroom cloud, aggregates and falls back to the earth as a shower of molten glass. The extant body of glass, trinitite, was the only material that remained at ground zero—The Gadget and the metal scaffolding holding the device were obliterated.

My interest in this peculiar place, a convergence point of time and the land, began with a fascination with trinitite. Physically speaking, trinitite is an unremarkable material. Composed primarily of a feldspar-rich arkosic sand and quartz particles, the resultant glass has little to offer the fields of mineralogy and geology. Trinitite has a pumice-like appearance, riddled with cavities and brittle, like dried bone. Many specimens have a greenish tint; rare pieces contain red flecks and streaks, remnants of the vaporized bomb casing and scaffold structure. It is typical of trinitite to have a smooth, glassy surface on one side, with a more rough and uneven texture on the opposite. The most prized examples are rounded and smooth on all sides, drops of trinitite believed to be the "rain" of glass falling back to the ground from the explosion, cooling to hardness before reaching the desert floor. Trinitite contains higher amounts of background radiation due to the cesium, barium, and other radiological elements contained within, but it is safe to handle with bare hands.

Trinitite questioned how glass could be used for containment and how a material is able to embody the complex history of its creation. I began to see Trinitite as a self-sufficient conceptual vessel, but it required interpretation and context. Despite the many layers of referent inherent to it, this information was not self-evident in its structure or appearance. If it is the duty of historians to accumulate, analyze and interpret moments in time, it can be said that trinitite demands the same strategy of classification, rather than existing as a mere mineralogical curiosity. In this train of thought, trinitite benefits from the authority of being viewed as an alternative artifact or reliquary. To move our understanding of the geologic forward, previous models of interpretation for history, geology, and the archive must be redefined, and composed into a new discipline of investigation. Trinitite, as a discreet object, provides the opportunity to view and understand the complex nature of events on a manifold level. Therefore, where the natural and academic sciences are frustrated, the realm of poetics intercedes.

As an artist, I felt that to merely visit this site was not enough. I wanted to formulate an experience that would be equal parts research endeavor, performance, and act of reverence. To that end, I chose to dress myself in the distinctive style of the lead scientist of the atom bomb's creation, Dr. J. Robert Oppenheimer. A genius of theoretical physics, Oppenheimer was an iconic, contrary figure for The Atomic Age and embodied the internal, moral struggles of the scientists involved with the construction of The Gadget. It was my hope that those who flocked to Trinity would recognize the costume; it would be as much an act of personal curiosity as ritual to conjure Oppenheimer's spirit. To see and be seen at Trinity was a condition of the research, an opportunity to address this transformed land as a living history.

Walking the Trinity Site, one has the strong feeling of a geographic point intersecting with geologic time. The very nature of nuclear technology, with the necessity to mine and refine naturally radioactive ores and minerals, is to engage with a timescale beyond human understanding. The half-lives of enriched minerals and isotopes spans into the tens of thousands of years, a source of harm to life for many hundreds of human generations. The threat of nuclear warfare and the mismanagement of nuclear stockpiles are not to be underestimated, nor minimized in any way. But perhaps the more insidious and prescient problems involve sites of radioactive contamination and waste accumulation. The process of creating enough uranium-238 and plutonium for the bomb required the creation of industrial facilities at Hanford, Washington and Oak Ridge, Tennessee. These sites produced massive amounts of volatile nuclear materials, and the ecological consequences of their production are a burden upon the present.

It is interesting to note that trinitite is not the only manmade glass containing radioactive substances and isotopes. The pioneering work done at Sellafield Limited in the United Kingdom has introduced an industry standard protocol to decommission highly radioactive substances by trapping the materials in a glass matrix, known as "vitrification." While this process allows the waste materials to become chemically inert, it does not diminish their radioactivity. Each batch of the formulated glass is poured into thick, stainless steel containers, themselves sheathed within thick concrete sleeves. Burial within a fortified repository is the most practical solution for how these containers will be dealt with, in the long term. But this is not a definitive solution, there are many problems that emerge and their solutions are ill defined. Due to the long half-life of industrial radioactive materials, the processed waste remains deadly to life for up to 50,000 years.

My experience at Trinity, combing the packed desert floor for bits of trinitite, offered an alternate interaction between the land and myself. I had formulated the trip as an examination of a strange place within the American landscape through my own artistic practice. Beyond the scattered trinitite and the surprising amount of spectators to be found at Trinity, I found an entry point to forces more basic and grand than myself. ■

—all images from *Trinity Pilgrimage*, Bryan M. Wilson, 2009

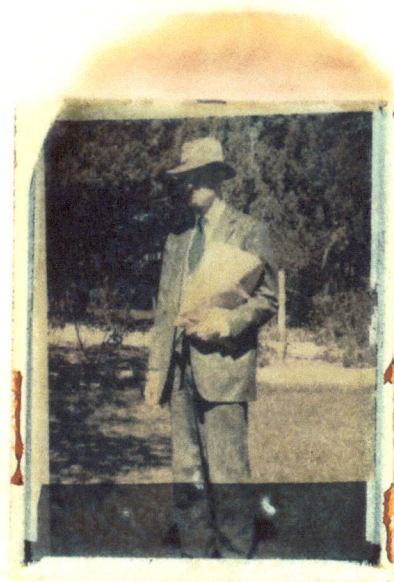

Geoff Manaugh + Nicola Twilley

## 38. ONE MILLION YEARS OF ISOLATION: AN INTERVIEW WITH ABRAHAM VAN LUIK[1]

Yucca Mountain, Nevada
Image courtesy of the US Department of Energy

*Abraham Van Luik is a geoscientist with the U.S. Department of Energy. At the time of this interview, in the autumn of 2009, Van Luik was based at the proposed nuclear waste-entombment site at Yucca Mountain, Nevada.*

*Yucca Mountain—an extinct supervolcano inside the Nevada Test and Training Range at the Nellis Air Force Base, 90 miles northwest of Las Vegas—is the controversial site originally chosen by Congress for the storage of nuclear waste. Its political fate remains uncertain. Although the Obama Administration has stated that Yucca Mountain is "no longer... an option for storing nuclear waste," Congress has since voted to continue funding the project—albeit only with enough funds to allow for the official licensing process to continue.*

*As part of a larger series of quarantine-themed interviews recorded and published from 2009-2010, we spoke to Abraham Van Luik about the technical nature of nuclear waste storage and what it means, on the level of geological engineering, to quarantine a hazardous material for more than one million years.*

**Geoff Manaugh:** How did you start designing a project like Yucca Mountain, when you're dealing with such enormous timescales and geological complexity?

**Abraham Van Luik:** You start with a question: How do you perceive the need to isolate a material from the environment?

---

1 Originally published on BLDGBLOG, 2 November 2009: http://bldgblog.blogspot.com/2009/11/million-years-of-isolation-interview.html.

I think most people would begin to answer that by looking at the nature of the material. Wherever that material is currently, we make sure that there is either a thick wall or a deep layer of water to protect the people working around it. That's what's being done at a reactor: when spent fuel comes out of a reactor, it's taken out remotely with no one present, and put into a water basin that's deep enough that there is no radioactive shine from the spent fuel escaping out of that water. If the pool is getting full, after five years or so of cooling, then the utility company will take the material out of the pool—remotely manipulated from behind leaded-glass windows—and put it into dry storage. Dry storage uses very thick steel and concrete. And there it will sit until someone disposes of it, or until it's reprocessed.

Now, in most countries, what they have done next is asked: What geology would be very good for isolating this material from the environment? And what geologies are available in our country? The Swedes have gone to their granites, because their whole country is basically underlain by granites. The French looked at granites, salts, and clay, and decided to go with clay. The Belgians and Dutch are looking at clay and salts; and the Germans are looking at salts right now, but also at granites and clay. The Swiss are looking at clay, mostly, although they did look at crystalline rock—meaning rock with large crystals, like granite, gabbros, and that kind of thing. But they decided that, in their particular instance—since the Alps are still growing and slopes are not all that stable over hundreds of thousands of years—to look instead at their deep basins of clays close to the Rhine River as a repository location. We're all looking to isolate this material for about a million years.

In the U.S., we did a sweep of the country, looked at all the available geologies, and we decided that we had *many* possible sites. We investigated some, which basically involved looking at what we knew from geological surveys of the states, and then we made a recommendation to go look at three of the possibilities in greater detail. There was then a decision process: it went from nine sites, to five, to three.

At that point, Congress stepped in. They started looking at the huge bills associated with site-specific studies—excavation is not cheap—and they said: let's just do one site and see if it's suitable. If it is not, then we'll go back and see what else we can do.

So that's how Yucca Mountain, basically, was selected. It was a cost-saving measure over the other two that were in the running for a repository. Those were a bedded salt site in Texas and a basalt site—a deep volcanic rock site—in Washington State.

But all three were looked at, and all three were judged to be equally safe for the first 10,000 years—which, at that time, was the regulation. Since the selection of Yucca Mountain, the regulation has been bumped up to a million years, which is pretty much where the rest of the world is looking: a million years of isolation.

Now, the reason that you want to isolate this material for a million years is that the spent fuel—meaning fuel that no longer supports the chain reaction that keeps reactors making electricity—contains actinides. These are metal elements, from 90 to 103 on the Periodic Table, most of which are heavier than uranium (which is 92). Actinides are generally very slow to radioactively decay into smaller atoms—which then decay more rapidly—and some of the actinides actually do remain hazardous for a million years and beyond. The trick is to isolate them for that length of time.

At Yucca Mountain, we took the attitude that, since we basically have a dry mountain in a dry area with very little rainfall, we would use a material that can stand up to oxygen being present. The material we selected was a metal alloy called Alloy 22. Our design involves basically wrapping the stainless steel packages, in which we would receive the spent fuel, in Alloy 22 and sticking them inside this mountain with a layer of air over the top. What we know is that when water moves through rock or fractured materials, it tends to stay in the rock rather than fall—unless that rock is saturated. Yucca Mountain is unsaturated, so water ought not be a major issue for us at Yucca Mountain—yet it is.

We have to worry about future climates, because, right now in Nevada, we are in a nine year drought—and, basically since the last Ice Age, we have been in a 10,000-year drought. 80% of the time, if we look a million years into the past, we have, on average, twice the precipitation we have now. Most of the past is—and the future will be—wetter and cooler. Which is nice for Nevada!

In any case, we tried to take advantage of the natural setting, as well as take advantage of a metal that stands up very well to oxidizing conditions. That is how, in our safety analyses, we showed that we are basically safe for well beyond a million years—*if* we do exactly what we said we would do in that analysis.

Other countries have decided not to go in a similar direction to us. The only other country that's contemplating a similar repository to ours is Mexico. All the other countries in the world are looking at constructing something that is very *deep*—and under the water table. If you go under the water table deep enough, there is no oxygen in the water, and if there is no oxygen than the solubility of a sizable number of the radionuclides is a non-problem. Many are just not soluble unless there is oxygen in the water.

Going that deep then allows those countries to use a different set of materials, ones that last a long time when there is no oxygen present. For example, the Swedes are using granite—so are the Finns, by the way, and the Canadians, though the Canadians might decide to go for clays. With the granites, the older they are, the more fractured they are, and they can't predict a million years into the future where the fracture zones are going to be. So they have chosen a copper container for their spent fuel; copper is thermodynamically stable in granite. In fact, copper deposits naturally occur in granite. They then wrap a very thick layer of bentonite clay around the container, which they put in dry. When that clay gets wet, as it will do eventually, it expands. When there is a fracture zone that is created by nature, the clay will basically decompress itself a little bit, fill the fracture zone, and you will still have a lot of protection from that clay layer. It's a similar set up with salt or clay repositories, they eventually close up against the waste packages. Nothing moves through clay or salt very rapidly.

Those are basically the three rock types that the whole world is looking at in terms of repositories.

So you can rely more on the engineered system or more on the natural system. Either way, it's the combination of the two systems that allows you to predict, with relative security, that you're going to isolate a material for well over a million years. By that time, the natural decay of the material that you've hidden away has pretty much taken care of most of the risk. In fact, by about half a million years, most of the spent fuel is less radioactive than the ore from which it was created. That's a wonderful argument—but the spent fuel still isn't safe at that point. You still need to continue to isolate it, just as you don't want to live on top of uranium ore, either. It's a dangerous material.

In a nutshell, that's our philosophy of containment.

**Manaugh:** I'm interested in how you go about testing these sorts of designs. Do you actually build scale models, like the U.S. Army Corps of Engineers' hydrological models, or do you rely on lab tests and computer simulations, given the timescale and complexity?

**Van Luik:** What we do is safety assessments that project safety out to a million years. What I used to say to my troops, when I was a manager of this activity, was: "Safety assessment without any underlying science is like a confession in church without a sin: without the one, you have nothing to say in the other."

Yucca Mountain, Nevada
Image courtesy of the US Department of Energy

To collect the science needed to make credible projections of system safety, we have dug several miles of tunnels under this mountain; we've done lots of testing of how water can move through this mountain, if there was more water; and we've done testing of coupons of the materials that we want to use. These tests were performed using solutions, temperature ranges, and oxygen concentrations that we think are representative over the whole range of what can be reasonably expected at Yucca Mountain. Those kinds of physical tests we have done.

We have also utilized information from people who have taken spent fuel apart in some of our national laboratories and subjected it to leaching tests to see how it dissolves, how fast it dissolves, and what dissolves out of it. We have done all of that kind of testing, and that's what forms the basis for our computer modeling.

One thing we have not done, and can't do, is a mock-up of Yucca Mountain. It just doesn't work that way. It's too complicated, too large, and too long a time-scale.

In compensation for that spatial and time-scale difficulty, what we have done is looked for similar localities with uranium deposits in them, like Peña Blanca, Mexico, just north of Chihuahua City. There, we have rock very similar to Yucca Mountain's rock, and we have probably a 30-million year old uranium deposit—quite a rich one—that was going to be mined until the price of uranium dropped considerably. We've studied that piece of real estate—it has roughly similar rock, sitting under similar conditions except for more summer rainfall—and we've looked at the movement of radioactivity from that ore body. From that we've gained confidence that our computer modeling can pretty much mimic what was seen at that uranium site.

We've looked for natural analogues of other possible conditions—for example, the climate at Yucca Mountain during an ice age. We've studied six or seven sites that mimic what we would see during a climate change here.

And, in terms of materials, there are some naturally occurring materials that have a passive coating on them. We've studied metals found in nature that are similar in the way they act to the metals that we are using for our waste packages.

So we have gone basically all through nature looking for analogous processes—but none are exact matches for Yucca Mountain. It's going to need something more unique than that. I think the same is true for every other repository being contemplated.

We have worked in cooperation with fourteen other countries through the European Commission's Research Directorate in Brussels, and the Nuclear Energy Agency in Paris, to compare notes on natural analogues and discuss what is useful and what is not for which concept. All these countries are doing the same kind of thing: looking at natural occurrences that are hundreds of thousands, if not hundreds of millions, of years old.

In some cases, the natural analogues we've studied are *billions* of years old. We've looked at the Oklo mining district in Gabon, Africa. We studied several occurrences in that mining district where, for the last few million years, ore bodies have been subjected to oxidizing conditions, because uplift of the land brought them above the water table. We've looked at these natural reactor zones, which were active two billion years ago when the earth was much more radioactive than it is now, to see what we could learn about the movement of radioactivity in an oxidizing zone. We can use that data for corroborating the modeling of Yucca Mountain.

On top of all that, we have the problem of unlikely volcanic events, as well as strong earth motions from equally unlikely seismic events, at Yucca Mountain. These are problems you won't have at most of the other repository sites being considered in the world. To study that, we brought in expert groups with their own insights and models to evaluate what the chances are, from a risk perspective, of a volcanic event actually interrupting or disrupting the repository. They also looked at the possibility of a very large ground motion adding stress and causing eventual failure of one or more of the waste packages. Although volcanic events are highly unlikely—as are very large ground-motion events—they must be factored into our analyses, based on the likelihood of their occurring over a one-million year time span.

We have basically done all safety-evaluation analyses from the perspective of the things that could happen, given the nature of this geologic setting. Looking at analogues for processes in nature has given us confidence that what we expect to see at Yucca Mountain is what we have seen nature produce elsewhere. These are indirect lines of evidence that support us—but we have also made a lot of direct measurements and observations, as well as testing in laboratories of materials and processes, to make sure that we're on the right track.

The National Academy of Sciences has reviewed our research and our situation, and they've agreed that we have predictability for about a million years. That judgment influenced the EPA, who then gave us a standard for a million years.

**Manaugh:** Could you discuss the material selection process in more detail? I'd like to hear how you found Alloy 22, for example. Also, when Nicola and I visited Yucca Mountain a few years ago, we were given a black glass bead at the information center. What role does that glass play in the containment design? Finally, are the materials you've chosen specifically engineered for the nuclear industry, or are these simply pre-existing materials that happen to have the requisite properties for nuclear containment?

**Van Luik:** No, the materials are not specifically engineered for the purpose of nuclear containment.

Let's look at Alloy 22 first. We looked at the whole range of what is commercially available in terms of pure metals and metal alloys. We also looked at things like ceramic coatings. There

Yucca Mountain, Nevada
Image courtesy of the US Department of Energy

are some very, very hard ceramic coatings that, for example, are used on bearings for locomotives. There are also ceramics that the military uses on projectiles to penetrate buildings. There are some very good ceramic materials out there, but we had a problem with the predictability of very, very long-term behavior in ceramics. That's why we decided to go with a metal. A metal will fail by several different corrosion mechanisms, but not by the breakage that is typical of ceramics.

One of the things that the metals industry has been doing—for the paper-pulp industry, for example, which creates the worst possible chemical environment you can imagine—is that they have been developing more and more corrosion-resistant materials. One of the top of the line of these corrosion-resistant materials was Alloy 22. We tested it alongside about six other candidates in experiments where we dripped water on them, we soaked them in water, and we had them half in and out of water, with varying solutions that we tailored for what we would expect in the mountain over time. The one that stood out—the most reliable in all of these tests—was Alloy 22.

The black glass that you saw is not something that the waste is wrapped in. This material will be made at Hanford and maybe at Idaho, too—and at Savannah River, they are making that black material right now. It's an imitation volcanic glass—a borosilicate glass—in which radioactive materials are dispersed. Material would be released from that if the waste package breaks, and if the material is touched by water or even water vapor. It would then start to alter, and as it alters it would start to release the radioactivity inside. So what you and your wife were looking at was basically a glass waste-form. We don't make it here—that's how radioactive waste will be delivered to us from the Defense Department and Department of Energy. We will receive it in huge containers, not as beads.

We also have little pellets of imitation spent fuel, similar to pencil lead in color, to show visitors what the fuel rods look like inside of a reactor. The fuel rods are ceramic, coated on the outside with an alloy.

**Nicola Twilley:** Could you walk us through the planned process of loading the waste into the mountain, all the way up to the day you close the outer door?

**Van Luik:** Sure. The process, depending on whether Yucca Mountain ever goes through, politically speaking, will be as follows.

From the cooling pools or dry storage at the reactor, we've asked the nuclear utility companies to put their spent fuel—or waste—into containers that we have designed and that we will supply to them. The waste will be remotely taken out of whatever container it is in now, put into our containers, which are certified for shipping as well as disposal, and then we would slide those containers onto trains. We want to use mostly trains—we try to avoid truck use.

Rail shipping containers currently in use are massive—some approaching two-hundred tons fully loaded. The trains would bring the containers to us and then we would up-end them remotely and take the material out in a large open bay—all done remotely, again. If it comes in the shipping cask that we have provided, we will be able to put it directly into the Alloy 22 and stainless steel waste package and weld it shut. Then, with a transporter vehicle that's also remotely operated, we would take it underground and place it end-on-end, lying down in our repository drifts. That's what we call the tunnels; tunnels without an opening are called *drifts*. We would basically fill the drifts until we get to the entrance, put a door on, and then move on to the next one. That's the basic scheme of how this would be done. Everything is shielded, of course, so that people are not exposed to radiation; workers are protected, as well as the public.

Yucca Mountain, Nevada
Image courtesy of the US Department of Energy

**Twilley:** How many containers could you fit inside a single drift, and how many drifts do you actually have in the mountain?

**Van Luik:** The drifts are each about 600 to 800 meters long. They vary a little bit, depending on where they are in the mountain. We will have 91 emplacement drifts -with an average of about 120 waste packages, set end-to-end, in each drift—to take care of the 70,000 metric tons that we are authorized to have. If we receive authorization to have *more* than 70,000 metric tonnes, then we're prepared to go up to 125,000 metric tonnes of heavy metal. That metric tonnage figure doesn't represent the total weight that goes into the mountain, by the way-it means that the containers have the equivalent of that many tonnes of uranium in them. In other words, 70,000 metric tons is about 11,000 containers that weigh about ten metric tons each, so it's a huge amount of weight. Each container contributes a significant amount of weight in itself: the steel and the Alloy 22 are very heavy.

In terms of what the repository would look like, if built, it would be a series of open tunnels, one after the other, with a bridging tunnel that allows the freight to be brought in on rail. Everything is done remotely. The 40km of tunnels would all be filled up at some point, and then we would seal up the larger openings to the exterior, but leave everything else inside the mountain unsealed.

This is very different, by the way, from every other repository in the world, which would tightly compact material around the waste packages. We want to leave air around the waste packages, because of our situation. We have unsaturated water flow, rather than saturated flow, and, as I've mentioned, water does not like to fall into air out of rock—it would rather stay in the rock, unless it's saturated and under some degree of pressure, such as from the weight of water above it. So if we put something like bentonite clay around our packages, that would actually wick the water from the rock toward the waste packages—which is a silly thing to do if you're trying to take advantage of an unsaturated condition.

**Twilley:** What process have you designed for sealing the exterior door? Does that also require a uniquely secure set of material and formal choices?

**Van Luik:** Sealing the repository wouldn't happen for at least 100 years, so what we have done at this point is basically left that decision for the future. We have done a preliminary design, which uses a heavy concrete mixture—as well as rock rubble for a certain portion—to seal the exits from the main tunnel that goes around and feeds all the smaller tunnels.

The idea is that these openings have nothing to do with how the mountain itself functions, because the mountain is a vertical-flow system. Coming in from the sides, as we are, has nothing to do with how the water behaves in the repository, or with the containment system we've designed. So we just want to block the side exits and make it very difficult for someone to reenter the mountain—to the point where they would basically be much better off reentering it by drilling a whole new entryway beside one of the old ones that's filled in.

Then there are going to be about seven vertical shafts for ventilation that will be sealed at the time of final closure. Those will be filled to mimic the hydrological properties of the rock around them; we don't want them to become preferred pathways of water, because those *will* point directly into the repository.

So there are two different closure schemes for the two different types of openings: three large entryways that will be completely sealed off to prevent reentry, and seven ventilation shafts that will be filled with materials that have been engineered to mimic the hydrological properties of the rock around it.

**Twilley:** And the ventilation shafts are required because the material is so hot?

**Van Luik:** Yes. Once we put the waste in, we want to blow air over it by drawing in air from the bottom and blowing it out the top to take heat away until we shut off the vents for final closure. The idea is to take enough heat out of the system so that, when we close it, it doesn't exceed our tolerances for temperature.

**Twilley:** Is there any chance that having such a large amount of heavy material at Yucca Mountain could actually pose a seismic risk for the region?

**Van Luik:** When we selected this particular location, we looked very carefully at faults. But you're right: if you get beyond a certain amount of weight, as under a growing mountain range, you do start shifting things in the ground. If you build something right on a fault line you can probably change the frequency of vibration at that location, and maybe aggravate the earthquake that's eventually going to happen.

However, even if we fill this repository to 125,000 metric tons, that is only something like .01% of the weight of the mountain itself. Plus, we are surrounded by two major faults, on both sides of the mountain, and even though there's movement occasionally on those faults, the block in the middle—where Yucca Mountain sits—is like a boat, riding very steadily. It's been like that for the past twelve million years, so we don't see that it's going to change in the future.

That said, we are in an area that's moving all the time. The entire area now is moving slowly to the northwest, and the basin and range here is still growing—the distance between Salt Lake and Sacramento is already twice what it was twelve million years ago, and they will continue to be pulled apart. We're well aware of the consequences of basin and range growth, and the possibility that the faults Yucca Mountain is sitting next to could be active again in the future. We factored that in. In fact, it's those earthquakes that might actually lead to failures in the system that would allow something to come out before a million years—otherwise, nothing would come out until beyond a million years.

But you can't put enough weight in that mountain to change the tectonic regime in the area.

**Manaugh:** Of course, once you have sealed the site, you face the challenge of keeping it away from future human contact. How does one mark this location as a place precisely not to come to, for very distant future generations?

**Van Luik:** We have looked very closely at what WIPP—the Waste Isolation Pilot Plant—is doing in New Mexico. They did a study with futurists and other people—sociologists and language specialists. They decided to come up with markers in seven languages, basically like a Rosetta Stone, with the idea that there will always be someone in the world who studies ancient languages, even 10,000 years from now, someone who will be able to resurrect what the meanings of these stelae are. They will basically say, "This is not a place of honor, don't dig here, this is not good material," etc.

What we have done is adapt that scheme to Yucca Mountain—but we have a different configuration. WIPP is on a flat surface, and their repository is very deep underground.; we're basically inside a mountain with no resources that anybody would want to go after. We will build large marker monuments, and also engrave these same types of warnings onto smaller pieces of rock and metal, and spread them around the area. When people pick them up, they will think, "*Oh*—let's not go underground here."

Now if people see these things and decide to go underground anyway, that becomes *advertent*, not *inadvertent*, intrusion—and we can't protect against that, because there's no way to control the future. All we're worried about is warning people so that, if they do take some action that's not in their best interest, they do so in the full knowledge of what they're getting into. The markers that we're trying to make will be massive, and they will be made of materials that will last a long time—but they're just at the preliminary stage right now.

What I have been lobbying for with the international agencies, like the International Atomic Energy Agency and the Nuclear Energy Agency, is that before anybody builds a repository, let's have world agreement on the basics of a marker system for everybody. Whoever runs the future, tens of thousands of years from now, shouldn't have to dig up one repository and see a completely different marker system somewhere else and then dig that up, too. They should be able to learn from one not to go to the others.

Of course, there's also a little bit of fun involved here: what is the dominant species going to be in 10,000 years? And can you really mark something for a million years?

What we have looked at, basically, is marking things for at least 10,000 years—and hopefully it will last even longer. And if this information is important to whatever societies are around at that time, if they have any intelligence at all, they will renew these monuments.

Yucca Mountain, Nevada
Image courtesy of the US Department of Energy

**Manaugh:** What kinds of projects might you work on after Yucca Mountain? In other words, could you apply your skills and a similar design process to different containment projects, such as carbon sequestration?

**Van Luik:** I think so—if we ever get serious about carbon sequestration. I don't know if you know this, but we laid off a lot of people here because there were budget cuts, and many of those people, because of the experience they had with modeling underground processes, are now working on carbon sequestration schemes for the energy sector and the Department of Energy.

No matter what happens to Yucca Mountain—whether it goes through or not—dealing with spent fuel and other nuclear waste will still be a problem, and that's the charter that was given to our office. What I'm hoping is that, as soon as Yucca Mountain gets completely killed or gets the go-ahead, I can go back to what I loved doing in the past, which was to look at selecting sites for future repositories.

One repository won't be enough for all time; it will be enough for maybe a hundred years, at the very most. You have to plan ahead. As long as you create the nuclear waste, you need to have a place to put it. Even if you reprocess it—even if you build fast reactors and basically burn the actinides into fission products so that they only have to be isolated for 500 years rather than a million—you still have to have a place to put that material. Even if we can build repositories less and less frequently, we will still be creating waste that needs to be isolated from the environment.

**Manaugh:** You mentioned that your favorite pastime was looking for repository sites. If you had the pick of the earth, is there a location that you genuinely think is perfect for these types of repositories, and where might that location be?

**Van Luik:** My ideal repository location has changed over time. When I worked on crystalline rock, like granites, I thought crystalline rock was the cat's meow. When I worked for a short time in salt, I thought salt was the perfect medium. Now that I have worked with the European countries and Japan for the past twenty-five years, learning of their studies of various repository locations, I'm beginning to think that claystone is probably the ideal medium.

In the U.S., I would go either to North or South Dakota and look for the Pierre Shale, where it grades into clay: there, you get the best of both worlds. I have been quoted by MSNBC, much to the chagrin of my bosses, saying that, if I were to get the pick of where we go next, that's where I would go. They really didn't like that—I was supposed to praise the Yucca Mountain site. But let's get real: Yucca Mountain was chosen by Congress. We have shown that it's safe, if we do what we say in terms of the engineered system. But it was not chosen to be the most optimal of all optimal sites; the site-comparison approach was taken off the table by Congress. As long as a chosen site and its system are safe, however, that is good enough.

Our predicted performance for Yucca Mountain, lined up to what the French are projecting for their repository in clay, and next to what the Swedes are projecting for their repository in granite, shows about the same outcome, over a million years, in terms of potential doses to a hypothetical individual. We're safe as anybody can be—which is what our charter requires. We told Congress in 2002 that, yes, it can it be done safely here—but it's going to cost you, and that cost is in Alloy 22 and stainless steel. Congress said OK and it became public law.

**Twilley:** Are any countries actually using their repositories yet?

**Van Luik:** They're getting very close to licensing in Finland and Sweden. Those are going to be the first two. We have a firm site selection in France, which means that they'll be going into licensing soon. Licensing takes several years in every country. In fact, we're in licensing now, except we had a change of administration and they've decided that they really don't want to do Yucca Mountain anymore. They want to do something else. They have every right to make those kind of policy decisions—so here we are.

No one is actually loading high-level waste or spent nuclear fuel into a repository yet. We have our own repository working with transuranic waste from the Defense program, in New

Mexico, and both the Swedes and the Finns have medium-level waste sites, which are basically geological disposal sites, that have been active for over a decade.

The Swedes and Finns have a lot of experience building repositories underground, and their situation is interesting. The Swedes are building a repository under the Baltic Sea, but in granites that they can get to from dry land. When there is a future climate change, however, there's going to be a period when the repository area will be farmable; it will be former ocean-bottom that is now on the surface. Their scenario is that, at the end of the next ice age, you might actually get a farmer who drills a water-well right above the repository.

The Finns actually have a very pragmatic attitude to this. They have regulations that basically cover the entire future span, out to a very long time period-but they also say that, once the ice has built up again and covered Finland, it won't be Finland. No one will live there. But it doesn't matter whether anyone lives there or not: you still have to provide a system that's safe for whoever's going to be there when the ice retreats.

We—as in the whole world—need to take these future scenarios quite seriously. And these are very interesting things to think about—things that, in normal industrial practice, you never even consider.

The repository program in England, meanwhile, went belly-up—because of regulatory issues, mostly—but it's coming back, and it's probably going to come back to exactly the same place as it was before. That's a sedimentary-metamorphosed hard-rock rock site at Sellafield, right by the production facility. No transportation will be involved, to speak of. That's not a bad idea, but they had to prove that the rock was good. The planning authority rejected their proposal the first time, so they dissolved the whole waste management company and now the government is going to take over the project; it's not going to be private anymore. In the end, the government takes over this kind of stuff in most places because the long-term implications go way beyond the lifetime of one corporation.

If there's any country that's setting a good example for waste disposal, it's Germany. They're the only country I know of who have the same kind of regulations for *hazardous* waste and *chemical* waste as they do for *nuclear* waste. There are two or three working geological repositories for chemical waste in Germany, and they have been working for a very long time. They're the only ones in the world. The chemical industry in the U.S. has basically said, no, no, don't go there! [*laughs*]

But I think Germany is right: if one thing needs to be isolated because it's dangerous, then the other thing—that *never* decays and is also dangerous—needs to be treated in the same way. The EPA does have a standard for deep-well injection of hazardous waste—they have a 10,000-year requirement for no return to the surface. That was comparable to what we had here, until the standard for Yucca Mountain got bumped up to a million years by Congress. But with some chemicals, regulations only require a few hundred years of isolation—that's all. Those things don't decay, so that doesn't make sense to me.

Anyway, I applaud Germany for their gumption—and they're very dependent on their chemical industry for income. It's not like they're trying to torpedo their industry. They're just saying: you have to do this *right*. ∎

# 39. TERMINAL ATOMIC: TECHNOGEOMORPHOLOGICAL MOUNDS

Mexican Hat Disposal Cell, UT.
Image Credit: CLUI Archive Photo, with aerial support by Lighthawk. 2012

Though the underground nuclear catacombs for America's spent nuclear fuel are yet to be created, radioactive tombs of America's various nuclear programs already exist today, with more to come. Most are repositories for the remains of uranium mills, processing facilities, weapons plants, and contaminated tailings, bulldozed into engineered isolation mounds designed to limit contact with their surroundings for hundreds of years. There are dozens of these mounds, across the country from Pennsylvania to Arizona, built mostly by the Department of Energy, and maintained by their Legacy Management office.

These disposal mounds are generally low, rectilinear piles with flat, sloping tops – terrestrial umbrellas, keeping moisture out of the pile as much as possible. In arid environments, the outer layer is a coating of coarse *riprap* rock, a dead space where nothing grows, where no soil forms, and no roots take hold that could pierce the radioactive core. This tough skin allows occasional rains to pass through it to the next layer, a low-permeability clayey mixture a few feet thick. Water drains off to the side of the pile through channels at the base held in place with more layers of crushed stone.

These disposal cells are located primarily in the Southwest, where natural uranium deposits were found and exploited. Some of these former uranium mills were set up secretly for the Manhattan Project. Most started in the 1950s, and many operated until the 1990s. Presently, only one conventional uranium mill is operating in the USA, the White Mesa Mill in Blanding, Utah, in the heart of the uranium district and Indian country. However, that may change as the nation shifts towards more self-reliant energy sources.

Each disposal cell covers many acres and as much as half a square mile. They resemble ancient pyramids or relics from a geometrical mound-building culture, like archeological forms made for the future. They represent the legacy of the most advanced technology of a global culture: the creation of the atomic bomb, the ability to destroy the world at the push of a button. They are part of the nationwide network of industrial sites created to extract, process, manufacture, and engineer nuclear fuel for reactors and weapons—a continent-wide landscape machine to concentrate a naturally occurring trace material into such compressed atomic density that it explodes with galactic energy.

These mound sites, byproducts of this effort, are the end of the line, meant to be unconnected to the rest of the world, like deadly anachronistic time capsules. These are the most negative of spaces, *non*places, meant to stay inert and isolated for as much of forever as possible, kept from the present, but destined for the future. ■

all images, CLUI Archive Photo
with aerial support by Lighthawk. 2012

Bluewater Disposal Cells, NM.

Grand Junction Disposal Cell, CO.

Gunnison Disposal Cell, CO.

Maybell Disposal Cells, CO.

Rifle Disposal Cell, CO.

Shiprock Disposal Cell, NM.

Slickrock Disposal Cell, CO.

Tuba City Disposal Cell, AZ.

Uravan Disposal Cell, CO.

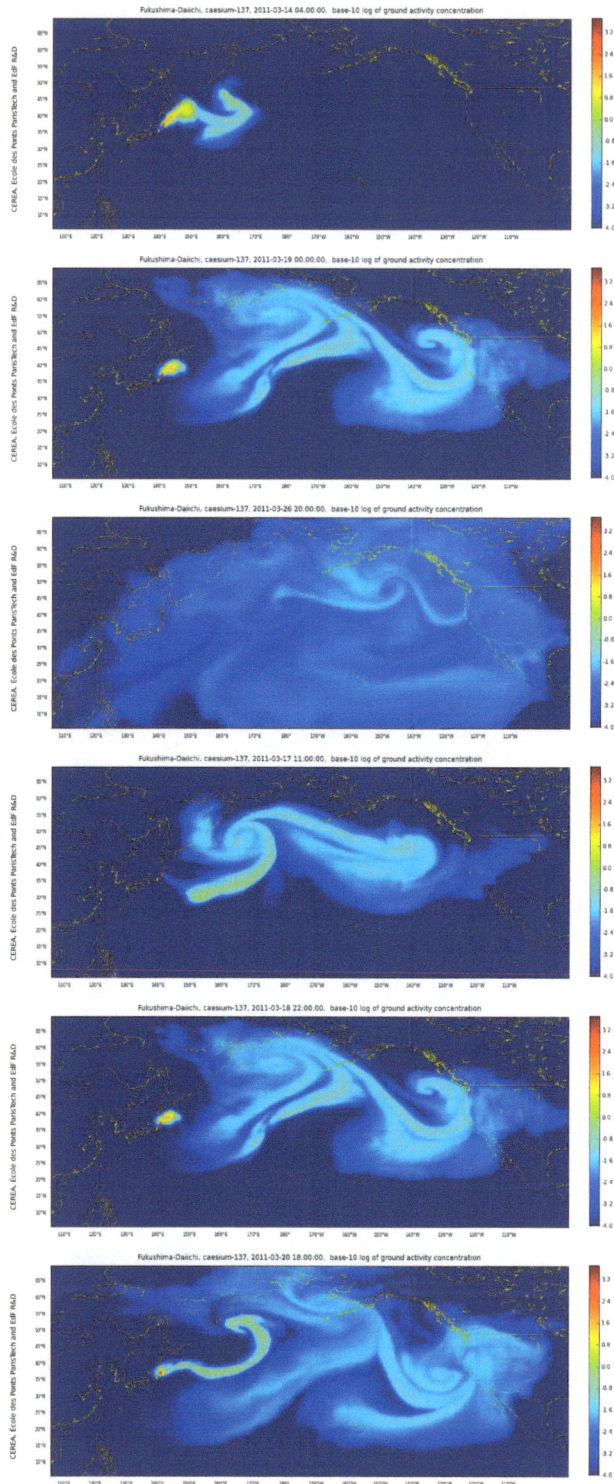

stills from CEREA's simulation of cesium-137 dispersion from Fukushima Daiichi from March 11-April 6, 2011 (image courtesy Centre d'Enseignement et de Recherche en Environnement Atmosphérique, http://cerea.enpc.fr/en/fukushima.html).

# AFTERWORD

Jane Bennett

## EARTHLING, NOW AND FOREVER?

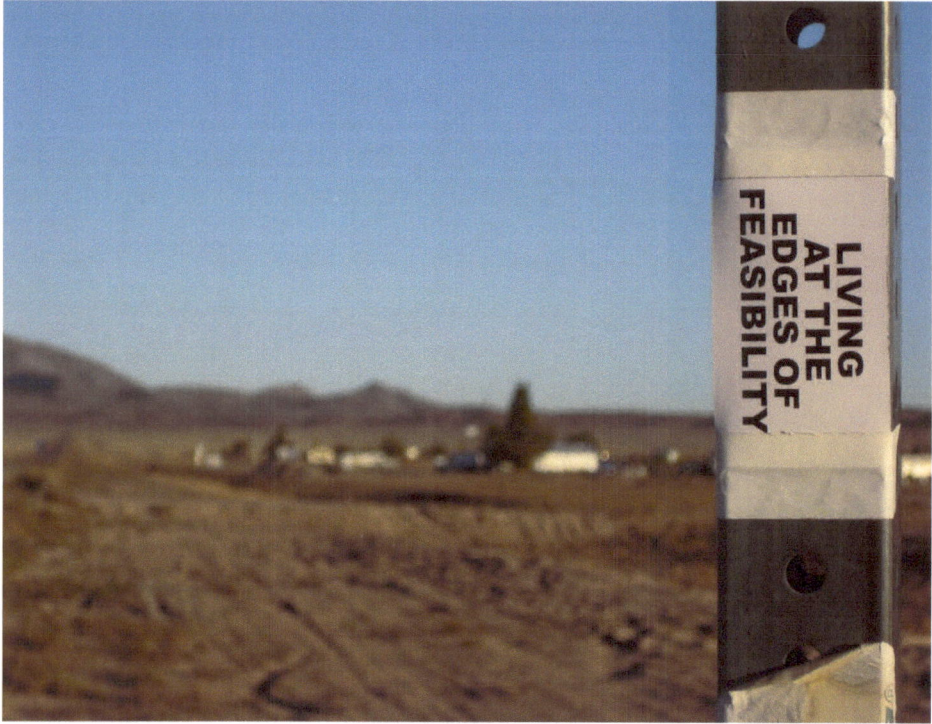

Ellsworth and Kruse have curated a powerful collection of words and images, "speculative or aesthetic devices" designed to instigate a new economy of discernment. It incites by showcasing a host of geologic actants—rivers, shale, escarpments, volcanos, tectonic plates, glaciers, river deltas, riprap, mud, clay—at work around and within human bodies, machines, cities, identities, architectures. Such "activations of geologic materiality"[1] help to recompose the default perceptual regime, which had tended to overlook the geologic (except on special occasions) and to forget how thoroughly we ourselves are geo-creatures—Earthlings.

We are Earthlings both in the sense that we need a host of other bodies ("the planet") to live and in the sense that "we" are made of the same elements as is the planet. We are "walking, talking minerals," redistributions of "oxygen, hydrogen, nitrogen, carbon, sulfur, phosphorous, and other elements of Earth's crust into two-legged, upright forms."[2] Like wind or river, human individuals and groups are geologic forces that can alter the planet in countless and, as the concept of the Anthropocene marks, game-changing ways.

Indeed, I think that one of the events that the idea of Anthropocene tries to capture is a certain *convergence* between two styles of temporality that we had formerly thought were distinct. The first style we had associated with the apersonal geologic: this was a bi-modal time

---

1  Elizabeth Ellsworth and Jamie Kruse, "Introduction: Evidence: Making a Geologic Turn in Cultural Awareness," in this volume.

2  Russian scientist Vladimir Ivanovich Vernadsky, quoted in Lynn Margulis and Dorion Sagan, *What is Life?* (Berkeley: University of California Press, 1995), 49.

of either a breakneck and explosively transformative speed (lightning, earthquake, wildfire) or an implacably slow, deep time (sedimentation, erosion, radioactive decay). The second style of temporality we identified as prototypically human: this was the more moderate, midrange speed of human endeavor, the enactment of intention or plan, the time of the everyday. There were, of course, exceptions, such as the superfast impact of the atomic bombs the Americans dropped on Hiroshima and Nagasaki, or (what many believed to be) the gradual accretion of human knowledge throughout history. The idea of the Anthropocene, however, suggests that the exceptions are not exceptional and that there is little reason to posit a hard, ontological dichotomy between human and ahuman forces, with regard to their temporalities or scope of effect.

Alongside a new economy of discernment, *Making the Geologic Now* also supports a different, more ecologically careful, economy of consumption and extraction. A keener cultural sensitivity to the deep and periodically explosive time of the geologic might, for example, encourage the pursuit of sources of energy that do not generate waste so toxic that it must be quarantined for one million years, or discourage practices like fracking that foul the water and might contribute to earthquakes. Affirming the geo-mode of long time also holds promise for lifting American political discourse above its currently idiocy, wherein crucial issues like climate change are elided for the sake of moralistic red herrings or theo-populist slogans. Geopolitical theorist Jairus Grove provides one example of this in a recent Facebook post: "Nothing says rearranging the deck chairs on the Titanic like two presidential candidates arguing over the size of tax cuts when fish are boiling to death in the Midwest.[3]

*Making the Geologic Now* hopes, as do I, that a sharper awareness of the geologic operative both around and through us would do us some (pragmatic, political) good. "The understanding of earth processes," write Ellsworth and Kruse, "can offer inspiration for how we might think about the qualitatively different ways that humans are now living on planet earth."[4] But they may put the point too mildly: Yes, this understanding can push us to "think" about modes of life, but this thinking is also a *judging* about which ways can best ensure our ongoing existence. It seems important to admit that at least one of the goals of the geologic turn is to figure out how to stick around as one species on the planet among numerous others—*how to maintain our Earthling status* in its various entanglements. For me, one of the effects of a heightened awareness of the interpenetration of the human and ahuman geologic is that it stretches my definition of "self"-interest to include the flourishing of the complex system of bio-geologic processes. This enriched understanding of "self" would then, I hope, enable a more extended pursuit of our conatus, the endeavor to persist in being.[5]

Bill Gilbert suggests that (alongside the pursuit of a new sensibility and a new political economy) we need a new "narrative as a nation and species," one that "encompasses the expanse of planetary time, not the fleeting moments of pop culture." We need this, he says, if we want the Anthropocene "to extend for thousands of years into the future." And we *do* want this extension, don't we? Don't we readers of *Making the Geologic Now* unconsciously project forward, if not the destructive, short-sighted version of the Anthropocene, a "cene" that includes, for as long as possible, the presence of the anthropos? When I make this implicit proj-

---

3  Grant Schulte, "Thousands of Fish Die as Midwest Streams Heat Up," *Associated Press: The Big Story*, 5 August 2012, http://bigstory.ap.org/article/thousands-fish-die-midwest-streams-heat.

4  Ellsworth and Kruse, "Introduction," in this volume.

5  Conatus is Spinoza's term, which Hasana Sharp explains this way: "all beings are provisionally individuated by their striving to persevere in being, and this endeavor to exist is their essence (*Ethics* III, p. 6). A human essence, or appetite to live, is called 'desire' insofar as we are conscious of it (III p9c)" (Hasana Sharp, *Spinoza and the Politics of Renaturalization* [Chicago: University of Chicago Press, 2011], 30-31).

ect overt, I see that I seek the postponement of the arrival of a radically posthuman future. And that this is a big part of why I (and perhaps other of the contributors to this collection) seek to open up "possibilities for humans to evolve ways to live in relation to geologic time."[6]

What of this conative desire, to be (or for there to be) Earthling(s) now and forever? Can the identification as Earthling be detached from the hope that Earth continues to include *us*? *Should* we try to detach geologic sensibility from *all* notions of self-interest? Is it really possible, given our current evolutionary form, to live according to the maxim that "while the human species can't get along without the geologic, the geologic will continue on in some form or other long after we have ceased being part of it?"[7]

I'm not sure of the answers to these questions, but if I had to come down somewhere, I'd say that the assumption of belonging and the tendency to project ourselves into the long-term horizon of the geologic run pretty deep. Or at least at this point in late-modern history they function as key parts of what *motivates* ecological practice and its various attempts to postpone the arrival of a posthuman earth and the vast suffering that it will be—aleady is—entailing. The idea of a deep belonging between human beings and a rather volatile earth also provides much of the energy for the political project called the geologic turn.

I liked Lars van Trier's 2012 film *Melancholia* for one reason: its realistic staging of the scene of the end of the earth (by means of a collision with a much larger planet) jolted me into remembering that, in addition to being a woman, an American, a teacher, a friend, on the Left, etc., I am also, perhaps first and foremost, an Earthling.[8] I live on and through an assemblage of materialites hurtling through space. I mention this because the will to belong to earth, to maintain Earthling status for the unforeseeable future, seems quite capable of persisting alongside a growing sense, within science, art, and popular culture, that this fit is nowhere guaranteed. The contemporary version of the will to belong is perhaps the latest in a long line of hopeful projections of a fittingness between humanity and the future. Religious versions include the notions of intelligent design and Providence. Focusing on the latter, Freud at first considered it an "illusion" susceptible to demystification, an infantile wish to be outgrown. But he too seems to have come ultimately to the conclusion that it is not so easy to vanquish—by reasoned analysis or the production of counter-narratives—the will to belong to the future. It is instead extremely resilient, born as it is "from man's need to make his helplessness [in the face of the overwhelming power of nature and death] tolerable."[9]

One doesn't have to be a theist (I'm not) to share something of the faith that humans belong to the earth, even if the earth doesn't belong to humans. Projecting ourselves into the geologic future may be what Earthlings do; pretending that this sort of projection will fulfill itself automatically is something we do too much. ∎

---

6 Ellsworth and Kruse, "Introduction," in this volume.

7 Ellsworth and Kruse, "Introduction," in this volume.

8 See also William Connolly, "Melancholia and Us," *The Contemporary Condition*, 26 May 2012: "The brilliance of *Melancholia* is that it ... allow[s] the experience of attachment [to Earth] to soak into our pores" (http://contemporarycondition.blogspot.com/2012/04/melancholia-and-us.html).

9 Sigmund Freud, *The Future of an Illusion*, ed. and trans. James Strachey (New York: W.W. Norton, 1961), 23-24.

# CONTRIBUTORS

## EDITORS:

### Elizabeth Ellsworth

is Professor of Media Studies at the New School, New York, and co-founder with Jamie Kruse of smudge studio, a nonprofit media arts and design collaboration. Her research and teaching focus on pedagogy as an expanded cultural practice capable of fostering new ways of thinking and knowing. In particular, she works with the idea that learning is an emergent phenomenon whose potentiality emanates from the disposition of things, processes, and people in pedagogical designs. She recently served as Associate Provost for Curriculum and Learning at The New School, and is author of *Places of Learning: Media, Architecture, Pedagogy* (Routledge, 2004) and *Teaching Positions: Difference, Pedagogy and the Power of Address* (Teachers College Press, 1997). Her recent journal articles focus on projects that fuse learning with aesthetic experience, and public pedagogy. Elizabeth earned her PhD in Communication Arts from the University of Wisconsin-Madison.

### Jamie Kruse

is an artist, designer and independent scholar. In 2006 she co-founded smudge, with Elizabeth Ellsworth, based in Brooklyn. She is the recipient of grants from the Graham Foundation for Advanced Studies in the Fine Arts; The New School Green Fund; New York State Council for the Arts and the Brooklyn Arts Council. Exhibitions include the Sheila C. Johnson Design Center; the Storefront for Art and Architecture; the Massachusetts Institute of Technology; Incident Report; and Richland Galleries. She has presented her work at Parsons, The New School of Design; the A. Alfred Taubman College of Architecture & Urban Planning, University of Michigan; the Center for Land Use Interpretation Los Angeles; the Oslo School of Architecture and Design; Thyssen-Bornemisza Museum, Madrid; and the California College of Arts. She is the author of *Friends of the Pleistocene*, fopnews.wordpress.com.

Web Designer:
**Alli Crandell**

Print Designers:
**Reg Beatty and Jamie Kruse**

———

## CONTRIBUTORS:

### Matt Baker

is an artist working in public space in Scotland. His work is highly specific to its location and often also intricately woven into a social context. He was awarded the Saltire Society Award for Art in Architecture in 2011 for projects completed in Scotland between 2009-11. For the last 13 years Baker has been placing experimental installations in the landscape of South West Scotland.

### Jarrod Beck

is an artist living and working in Brooklyn, NY. He has created installations for the Lower Manhattan Cultural Council (NY), Lawndale Art Center (TX), MASS Projects (TX), Backroom (Madrid, Spain), Generator Projects (NM), and the Instituto Cervantes (NY). He has been a resident artist at the Lower East Side Printshop, Frank Lloyd Wright School of Architecture, Robert Blackburn Printmaking Workshop, Fine Arts Work Center and Socrates Sculpture Park. Beck's drawings are included in the Judith Rothschild collection of contemporary drawings at the Museum of Modern Art, New York. He earned an MFA from the University of Texas at Austin and a Master of Architecture degree from Tulane University.

### Stephen Becker

is a development consultant and unlicensed architect in Massachusetts. He is co-founder of mammoth, an architectural research and design collaborative, and a founding member of the ExEx. His work has been published in various magazines and journals including *Bracket*, *MONU*, and *Crit*.

### Brooke Belisle

received a master's degree from NYU's Interactive Telecommunications Program and a PhD from UC Berkeley in Rhetoric, Film, and New Media. Her research and teaching connects contemporary media art with ideas and formats from earlier moments in visual culture. She is currently writing on photographers, filmmakers, and digital artists whose work invokes the panoramic, the stereoscopic, and the history of astronomical imaging.

### Jane Bennett

is professor of political theory at Johns Hopkins University. Her latest book is *Vibrant Matter: A Political Ecology of Things* (Duke University Press, 2010). She is also the author of *The Enchantment of Modern Life and Thoreau's Nature: Ethics, Politics, and the Wild*. She is currently working on a study of Walt Whitman's materialism.

### David Benqué

Is a designer and researcher working in London, UK. He holds an MA from the Design Interactions department at the Royal College of Art and a BA in graphic and typographic design from the Royal Academy of Arts in the Hague, Netherlands.

### The Canary Project/ Susannah Sayler and Edward Morris

Susannah Sayler and Edward Morris work with photography, video, writing and installation. Of primary concern are contemporary efforts to develop ecological consciousness and the possibilities for art within a social activist praxis. In 2006 they co-founded The Canary Project—a collaborative that produces visual media and artworks that deepen public understanding of climate change.

### The Center for Land Use Interpretation

is an educational organization established in 1994 to increase and diffuse information about how the nation's landscape is apportioned, utilized, and perceived, and to help people become more aware of the physical characteristics and cultural significance of the shared landscape of the nation.

### Brian Davis

studied landscape architecture in North Carolina and Virginia and has practiced in Raleigh, Buenos Aires and New York City. He writes the landscape blog faslanyc and contributes to design journals focusing on projects and ideas related to landscapes of Latin America, New York City, and Appalachia. He is also a founding member of the ExEx research collaborative.

### Seth Denizen

works as a landscape architect and researcher at the University of Hong Kong. He graduated in 2007 from McGill University where he studied the Pliocene evolutionary biology of the Panamanian Isthmus. His work in art and architecture has engaged with the aesthetics of scientific representation, madness and public parks, legal geomorphology, and the political economy of construction waste.

### Anthony Easton
is interested in western culture, (dis)embodied theology, disability studies, and vernacular forms; he's shown in galleries, libraries, churches, bathhouses, and an accounting office in Poughkepsie. One of his artist books is in the library of the National Gallery of Canada, one of his essays has been anthologized by Routledge.

### Valeria Federighi
is a licensed architect. She holds a MArch from the Politecnico di Torino, with a thesis on incremental design in the slum of Dharavi, Mumbai; and a MS in Design Research from University of Michigan with a research on the photographic practice of Ruin Porn and its reflection in the morphology of Detroit. Valeria's work experience includes internships at AndersonAndersonArchitecture, San Francisco; URBZ, Mumbai; and Isolarchitetti, Torino.

### William L. Fox
Director of the Center for Art + Environment at the Nevada Museum of Art, is an art critic, science writer, and cultural geographer. He is a fellow of the Royal Geographical Society and recipient of fellowships from the Guggenheim Foundation, National Endowment for the Humanities, and National Science Foundation.

### David Gersten
is an architect, writer and educator based in New York City. He is a Professor in the Irwin S. Chanin School of Architecture at the Cooper Union, where he holds the Feltman Chair and has served as Associate Dean under Dean John Hejduk as well as Acting Dean of the School of Architecture. He has been a visiting professor, taught workshops and lectured in universities throughout Europe and the Americas. His drawings and stories have appeared in international exhibitions, and are held in the collection of the Canadian Center for Architecture, the NYC Public Library' print collection and numerous private collections.

### Bill Gilbert
has served on the faculty in the Department of Art and Art History at the University of New Mexico since 1988 where he holds the Lannan Endowed Chair as founder of the original Land Arts of the American West program. Gilbert is also the co-founder with Basia Irland of the new Art & Ecology emphasis in studio art and has recently been appointed as Acting Dean of the College of Fine Arts.

### Oliver Goodhall
holds a MA from the Royal College of Art in Design Interactions, having previously graduated from the Bartlett School of Architecture in 2005, and co-founded the architecture practice *We Made That*. He is interested in developing projects that expand engagement between strategic thinking and creative practice in the public realm.

### John Gordon
is Earth Science Policy & Advice Manager with Scottish Natural Heritage inEdinburgh, and an Honorary Professor in the School of Geography and Geosciences at St Andrews University. He is also a glacial geologist and has written extensively about the landforms and glacial history of Scotland, as well many other areas of the world, including the Arctic and the Antarctic. In 2003, he was co-leader of the Royal Scottish Geographical Society's Scotia Centenary Expedition to South Georgia in the Antarctic.

### Ilana Halperin

is an artist, originally from New York, and currently based in Glasgow, Scotland. She has a deep love of geology and shares her birthday with the Eldfell volcano in Iceland. Her work explores the relationship between geological phenomena and daily life.

### Lisa Hirmer

is an artist, designer and writer. She has a Bachelors of Architectural Studies (2005) and a Masters of Architecture (2009) from the University of Waterloo and is a founding principal (along with Andrew Hunter) of DodoLab, an experimental program that researches, engages and responds to contemporary community challenges, with a particular focus on the natural world, social systems, the built environment and cities in transition. She is based in Guelph, Ontario, Canada.

### Rob Holmes

lives, teaches, and practices as a landscape architect in Virginia. He is co-founderof mammoth, an architectural research and design collaborative, and a founding member of both the ExEx and the Dredge Research Collaborative. His work has been published in various magazines and journals including *Bracket*, *Urban Design Review*, *Landscape Architecture*, *MONU*, and *Crit*.

### Katie Holten

grew up in rural Ireland and lives in New York City. Her work explores the relationship between human beings and the environment. At the root of her practice is a curiosity about life's systems— both organic and man-made. In 2003 she represented Ireland at the Venice Biennale and in 2009 created *Tree Museum*, a public artwork celebrating the centennial of the Grand Concourse in the Bronx. She has had solo museum exhibitions at the New Orleans Museum of Art (2012), the Hugh Lane, Dublin (2010), the Bronx Museum, New York (2009), the Nevada Museum of Art, Reno (2008), the Villa Merkel, Esslingen (2008), and the Contemporary Art Museum St. Louis (2007).

### Jane Hutton

is a landscape architect and assistant professor in landscape architecture at the Harvard Graduate School of Design. Her work focuses on the externalities of material practice in landscape architecture, looking at links between the landscapes of production and consumption of common construction materials. In 2010, she curated the exhibition, *Erratics: A Genealogy of Rock Landscape*, looking at the cultural and scientific antecedents of rock-focused landscape architecture projects. Hutton is a founding editor of the journal *Scapegoat: Architecture, Landscape, Political Economy*, and is co-editor of Issues: 01 *Service* and 02 *Materialism*.

### Julia Kagan

is a science and health journalist. The former editor of *Consumer Reports* and *Psychology Today*, she is writing a book about New York City earthquakes. She is a former Visiting J. Stewart Riley Professor at the Ernie Pyle School of Journalism at Indiana University. A graduate of Bryn Mawr College, she is currently finishing an MFA in creative nonfiction at Bennington Writing Seminars.

### Wade Kavanaugh and Stephen B. Nguyen

have collaborated on large scale art installations since 2005. Although they each maintain their individual studio practices, their collaborative artworks have allowed them the freedom to investigate and juxtapose phenomena from the natural and built environments. Their works range from large sculptural objects to warehouse-sized immersive environments that suggest layers of earth, old growth forests, or the flow of a glacier. Their work has been exhibited across the United States, most recently at Mass MoCA in North Adams, MA, Carnegie Mellon University in Pittsburgh, PA, and at the Sun Valley Center for the Arts in Ketchum, Idaho.

### Oliver Kellhammer

is a Canadian land artist, permaculture teacher, activist and writer. His botanical interventions and public art projects demonstrate nature's surprising ability to recover from damage. His work facilitates the processes of environmental regeneration by engaging the botanical and socio-political underpinnings of the landscape, taking such forms as small-scale urban eco-forestry, inner city community agriculture and the restoration of eroded railway ravines. His process is essentially anti-monumental - as his interventions integrate into the ecological and cultural communities that form around them, his role as artist becomes increasingly obscured. He describes what he does as a kind of catalytic model-making, which lives on as a vehicle for community empowerment while demonstrating methods of positive engagement with the global environmental crisis.

### Elizabeth Kolbert

has been a staff writer at The New Yorker since 1999. Her three-part series on global warming, "The Climate of Man," won the 2006 National Magazine Award for Public Interest, the 2005 American Association for the Advancement of Science Journalism Award, and the 2006 National Academies Communication Award. Kolbert came to the magazine from *The New York Times*, where she wrote the Metro Matters column and, from 1992 to 1997, was a political and media reporter. Her first book, "The Prophet of Love: And Other Tales of Power and Deceit," was published in 2004. Her second book, "Field Notes from a Catastrophe," (2006), on global warming, is now available in paperback. Kolbert lives in Massachusetts.

### William Lamson

is a Brooklyn based artist who works in video, photography, performance and sculpture. His work is in the collections of the Brooklyn Museum, the Dallas Museum of Art, the Museum of Fine Arts in Houston and a number of private collections. Since graduating from the Bard MFA program in 2006, his work has been shown at The Indianapolis Museum of Art, The Brooklyn Museum, P.S.1 MOMA, and the Museum of Fine Arts in Santa Fe, among others. He recently completed two site-specific installations for Storm King Art Center.

### Janike Kampevold Larsen

is a postdoctoral fellow at the Oslo School of Architecture and Design. Originally a literary scholar, she is part of a research project called Routes, Roads and Landscapes, Aesthetic Practices en route, 1750-2015. She is working on a book called *Post National Natures* and co-edited *Routes, Roads and Landscapes* (Ashgate, 2011).

### Tim Maly

writes about cyborgs, architects, and our weird broken future at Quiet Babylon. He's the project coordinator of Upper Toronto, a science fiction design proposal to build a new city in the sky and Border Town, an independent design studio about divided cities. He is a co-founder of the Dredge Research Collaborative. He's part of the *Wired* Design staff and his work has appeared in *The Third Coast Atlas*, *McSweeney's*, *Icon*, *The Atlantic*, and *Volume* Magazine. He lives in Toronto.

### Geoff Manaugh

is the author of BLDGBLOG, former senior editor of Dwell magazine, and a contributing editor at *Wired* UK. In addition to regular freelance work for such publications as *Volume*, *Popular Science*, *The New York Times*, and *Icon*, he also co-directs Studio-X NYC, an off-campus event space and urban futures think tank run by the architecture department at Columbia University. In 2009-2010, with Nicola Twilley, Manaugh organized and curated an independent design studio and exhibition called "Landscapes of Quarantine" at New York's Storefront for Art and Architecture.

### Don McKay

is an award-winning Canadian poet, editor, and educator. Born in Owen Sound, Ontario and raised in Cornwall, McKay was educated at the University of Western Ontario and the University of Wales, where he earned his PhD in 1971. He taught creative writing and English for 27 years in universities including the University of Western Ontario and the University of New Brunswick. McKay is the author of twelve books of poetry, including *Long Sault* (1975), *Lependu* (1978), and *Apparatus* (1997). He has twice won the Governor General's Award, for *Night Field* (1991) and *Another Gravity* (2000). In June 2007, he won the Griffin Poetry Prize for *Strike/Slip* (2006).

### Rachel E. McRae

lives between Los Angeles and Toronto. She has exhibited in the USA and Canada; at events such as Toronto's Nuit Blanche, the New York Book Arts Fair and Art Basel Miami Beach; and curated programs with the Inside Out GLBT Film and Video Festival and Pleasure Dome.

### Brett Milligan

is the director and author of Free Association Design (F.A.D.) and a founding member of the ExEx research collaborative. He practices landscape architecture in Portland and teaches design courses at the University of Oregon. His work has been exhibited internationally and his writings have appeared in publications such as *MONU*, *Bracket*, and *The Journal of Landscape Architecture*. His current research on the infrastructure of the Klamath River Basin is funded by a grant from the Graham Foundation.

### Christian MilNeil

is a freelance writer, programmer, and author of *Vigorous North*. He lives in Portland, Maine.

### Laura Moriarty

is a painter, sculptor and printmaker whose works explore geologic time. Laura's work is exhibited widely, and she has participated in many guest lectures and residencies nationally and internationally, including the Ucross Foundation, Women's Studio Workshop, the Frans Masereel Center in Belgium, and the University of Virginia, Charlottesville. Her honors include two Pollock-Krasner Foundation Grants, along with grants from the New York Foundation for the Arts and the Crafts Alliance of New York State.

### Erika Osborne

is an artist and professor at West Virginia University. Her nationally exhibited work uses various media to address cultural connections to the environment. Erika's creative interests have crossed over into her teaching practice, driving her to develop two field-based courses for artists, Place: Appalachia and Art and Environment.

### Trevor Paglen's

work rejects the formal and disciplinary boundaries between art, social science, journalism, and other forms of intellectual and cultural production. He became an artist at age six, studied Religious Studies and Music Composition as an undergraduate at U.C. Berkeley, earned an M.F.A. from the School of the Art Institute of Chicago, and went on to complete a PhD in human geography from U.C. Berkeley. He lives and works in New York City.

### Anne Reeve

received her MA in the History of Art from University College London. She currently serves as Curatorial Research Assistant at Glenstone, where she produces filmed Oral History interviews for the Institution's research archive. Her writing has appeared in Art in America Magazine.

### Chris Rose

is a designer and researcher in knowledge building in the field of Design. He currently teaches in Furniture Design and in Graduate Studies at RISD. He has taught and organized specialized seminars in design thinking, creative process and materials in Finland, India, Italy, France, Holland, UK and USA. He is a contributor to the Engineering Social Justice and Peace (ESJP) network, and to a number of research publications including the Victoria and Albert Museum and the University of Brighton, UK. He is currently an Arts-Science collaborator at RISD under an NSF research program.

### Victoria Sambunaris

received her MFA from Yale University in 1999. Each year, she structures her life around a photographic journey crossing the American landscape. Her most recent project has been following the US/Mexican border photographing the intersection of geology, politics and culture along the volatile international boundary.

### Paul Lloyd Sargent

is multidisciplinary artist currently in the PhD program in the Department of Media Study at the University at Buffalo. His research investigates the legacies of the supply and disposal chains, primarily focused on the impact of the international shipping industry on ecologies, economies, and communities connected by the Great Lakes and St. Lawrence River.

### Rachel Sussman

is an artist and photographer based in Brooklyn. Her critically acclaimed project *The Oldest Living Things in the World* spans disciplines, continents and millenia. Starting at "year zero" and looking back from there, Sussman is creating an original index of continuously living organisms 2000 years old and older. She's spoken at TED, The Long Now Foundation, CNN and the BBC, and has received numerous awards and international press. Her work has been exhibited in museums and galleries in Europe and across the US.

### Shimpei Takeda

was born in Fukushima, Japan. He is a Brooklyn-based artist working with photographic materials. He primarily focuses on cameraless photographic techniques to capture otherwise unseen interactions of materials and light. As the Fukushima Daiichi nuclear disaster occurred close to where his family resides, within 40 miles, Takeda began working on an on-going project, "Trace—cameraless records of radioactive contamination."

### Chris Taylor

is an architect, educator, and director of Land Arts of the American West at Texas Tech University where he teaches in the College of Architecture. Since 2001 he has been developing Land Arts as a semester abroad in our own backyard that investigates the intersection of geomorphology and human construction. The books *Land Arts of the American West* and *Incubo Atacama Lab* document his field based investigations of multivalent landscapes. Taylor is a graduate of the University of Florida, the Graduate School of Design at Harvard, and recipient of the Steedman Fellowship from Washington University in St. Louis.

### Ryan Thompson

is based in Chicago, IL where he is an Assistant Professor of Art and Design at Trinity Christian College. His ongoing 'Department of Natural History' projects engage the complex and often strange relationships humans produce in collaboration with natural phenomenon.

### Etienne Turpin

teaches architecture theory and design research at the Taubman College of Architecture & Urban Planning, University of Michigan. In collaboration with Meredith Miller, he is principal investigator for the research initiative *Architecture + Adaptation: Designing for Hypercomplexity*, which analyses and provokes the agency of architecture in relation to extreme conditions of the Anthropocene. Etienne is a founding editor of the architecture, landscape, and political economy journal *Scapegoat* and the editor of *Architecture in the Anthropocene: Repositioning Design Research* (MAP Books Publishers).

### Nicola Twilley

is the author of *Edible Geography*, cofounder of the "Foodprint Project" with Sarah Rich, former Food Editor of *GOOD* magazine, and co-director of Studio-X NYC. Nicola is also curator of a forthcoming exhibition at the Center for Land Use Interpretation in Los Angeles exploring North America's spaces of artificial refrigeration; she is working on a book on the same topic. In June 2012, Future Plural—a curatorial and publishing initiative run with Geoff Manaugh—launched *Venue*, a pop-up interview studio and mobile media rig traveling around North America through September 30, 2013.

### Bryan M. Wilson

is a visual artist investigating themes of time, identity, and the body through a variety of craft and art disciplines. Nationally and internationally exhibited, his work takes the forms of objects, video, installation and image. He currently lives and works in New York City.

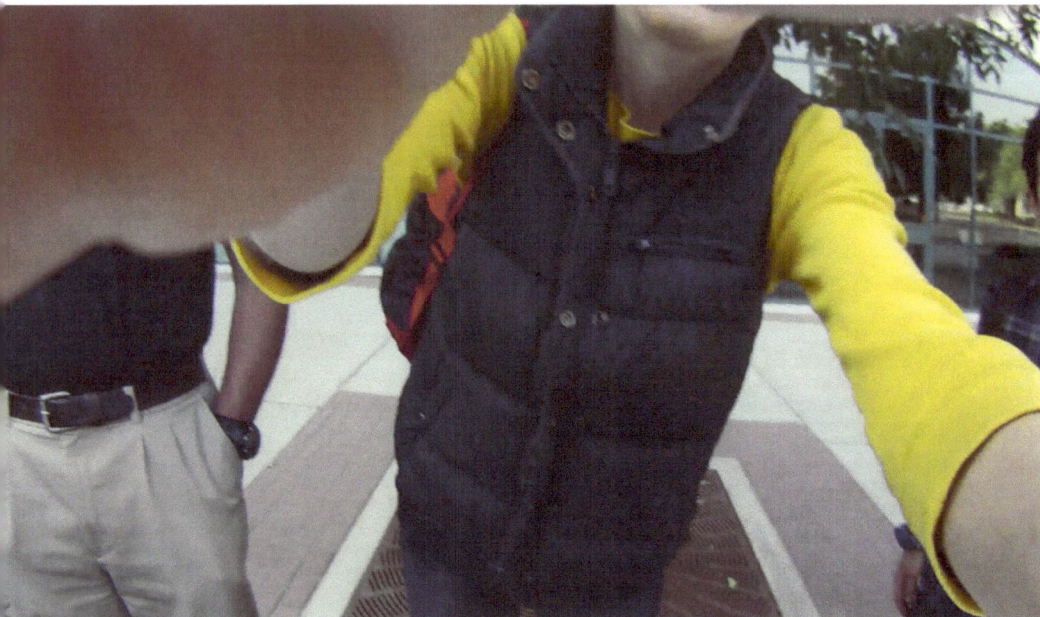

setting up the car-mounted video camera outside the EnergySolutions office in downtown Salt Lake City, smudge studio 2012

**MEETING SITES AND MOMENTS WHERE THE HUMAN AND THE GEOLOGIC CONVERGE**
SMUDGESTUDIO.ORG

www.ingramcontent.com/pod-product-compliance
Lightning Source LLC
Chambersburg PA
CBHW061138030426

42334CB00004B/82